# 景觀設計實務

## TOWNSCAPE DESIGN
## IN EUROPE

井上　裕
*YUTAKA INOUE*

●

写真/井上裕・浩子
*Photographs by Yutaka & Hiroko Inoue*

# 序
## FOREWORD

　　本書蒐集了我在1989年3月起至7月間的旅行和在1990年、1992年、1993年、1994年4月至5月間的短期旅程中，共歷時六個月左右間所拍攝而成的，大約有一萬張的歐洲鄉村幻燈片中，列舉各城鎮的景觀要素而整理作成的。

　　在歐洲，其實多半是美麗的城鎮；我們去探求到的這份美麗和愉快的秘訣，其實僅只是皮毛而已。但是再一想，由本書所介紹的照片亦可以清楚地明白，這份美麗和快樂的秘方，是和各城鎮的成長過程，風土歷史等長期以來的傳統，有密不可分的關係。

　　只是，也決非全然是風土歷史，傳統文化的影響。在現代文明之中，非但不破壞傳統城鎮之美，反而加以保護提升其重要性，是當今歐洲積極努力的課題。在本書後半段的內容裡，將加以解說目前歐洲方面所下的工夫和努力的方向。

*This book is based on research done during travels through Europe, beginning with a trip which lasted from March to July, 1989, and including three shorter trips made between April and May of 1992, 1993, and 1994. During these six months over ten thousand slides were taken of villages and towns in every part of Europe, and these photographs have been used to illustrate the architecture and various elements of these villages and towns.*

*There are truly a great number of beautiful villages and towns in Europe. In spite of our attempts to wish to experience the beauties and pleasures of all these places we were only able to visit a fraction of them. Nonetheless, as the photos in this volume show, the secret of their beauty was revealed to us in distinctive natural features and histories born in each.*

*However, not everything can be attributed to such factors as setting, history, tradition and culture. Europeans, thanks to their modern civilization, have recognized that it is not enough merely to preserve the beauty and pleasure of their villages and towns, but have seen the necessity of going further, and actually taking action to improve and develop this heritage.*

# CONTENTS

# SETTING

● 佈景

　　若說房屋城鎮的內部景緻，相對的，城鎮外圍的景觀（包含周圍的自然景觀）就是房屋的佈景。因此有時我們走進景觀壯麗的城鎮時，卻發現其中的房舍令人失望，相反的，有些城鎮街景美觀，房舍整齊但外圍景觀卻枯燥無味，缺乏吸引力。

　　對小村莊、城鎮而言佈景在整體設計中，佔有極重要的地位。由於小城鎮大都市不同，小城鎮可藉由地形及視野將周圍的自然景緻融入其中，盡收眼底。因此這種涵蓋自然景觀的小城鎮給人的印象及其受矚目的程度往往高過其自然景觀。

　　事實上，有許多舊式的小城鎮因其優越的佈景而成為有名的觀光勝地。

While the streetfront is related to the interior view of a village or town, the setting relates to the view of the village or town from the outside, and includes its natural surroundings. Thus it is sometimes the case that villages and towns with wonderful settings appear quite drab from within, or that villages and towns with beautiful streets are located in poor settings.

However, among the design elements the setting of villages or small towns is of special importance. Unlike large towns and cities, the natural setting of villages and small towns can be taken in together with the settlement in its entirety, depending on the viewpoint and the terrain, and thus the natural setting tends to make a strong impression on viewers. The result is that there are many examples of villages and small towns which have become famous tourist attractions since olden times, because of their impressive settings.

1．克洛威（英國，蘇格蘭）

# 水邊
Waterfront

　　海、河川、運河、湖等所謂水邊的佈景 (Setting)，也賦予小城鎮重要的特徵。其他地方沿著廣場或街道的房屋，在此成爲沿水而居的景色。不管是靜止的水或是潺潺流動的流水，大多是較地面略低，而僅有表面的波紋汕漾。這些波紋會因水面上的微風而變化，使倒映於水中的影像隨之搖擺，震動。不只是號稱世界最美的水都威尼斯，凡是傍水而居的住家，都可以看見水面美麗的倒影。

Sea coasts and rivers, lakes and canals are waterfronts which lend a special feature to villages and small towns. Instead of viewing houses situated along a streetfront or square one is seeing houses built along a waterfront. Quiet, slowly flowing water is always at a lower elevation than the land, its surface moving in the form of waves. While changing in response to the strength of the wind, waves reflect the special features of the town, distorting and blurring them. Not only Venice, said to be the world's most beautiful city, but all towns with streetfronts forming a background to the water, or reflected upon the water, appear beautiful to the eye.

2. 被稱爲 "小威尼斯" 的一角（法國，阿爾薩斯）

3. 歐魯那最古老的地區的一角。家家戶戶背對著盧河；同時這個小城也因是名畫家古斯達夫・庫魯貝（1819-77）的出生地而聞名。（法國，法蘭西・康德）

1. Crovie, Scotland, Great Britain.
2. The "little Venice" in Colmar, Alsace, France.
3. In the oldest part of Ornans, houses face straight back onto the river Loue. The town is the birthplace of painter Gustave Courbet (1819-77). Franche-Comptè, France.

4. 埃米安大教堂背面運河沿岸溫馨的住家。（法國，皮卡爾地）

5. 沿高黑爾河岸的木造建築。施韋比施哈爾德國，霍恩洛亞

6. 沐浴在夕陽餘光中的美麗街影映入運河中；但左邊的大型現代建築物稍稍破壞了
這幅美景。肯特（比利時，佛蘭德斯）

7. 運河沿岸的樹木遮住了房舍，故得以看到靜謐的田園景觀，但背面仍隱藏著不少
華麗的店舖。這是在距離比利時邊境3km處的斯露易斯的小鎮。（荷蘭）

8. 石灰岩山的峽谷河口形成的城鎮，河岸對面所見之景緻。奧米休（舊南斯拉夫，
達爾馬提亞）

9. 里亞爾橋所見的大運河沿岸街市。威尼斯（義大利）

10. 蘇伊特河岸遠眺必皮布爾斯鎮（蘇格蘭波坦地區）

11. 和薩林河相接的弗里堡的古老街景之一。（瑞士）

4. Charming houses along the canal behind the Amiens cathedral. Picardie, France.
5. Timber-framed houses looking onto the river Kocher. Schwäbish Hall. Hohenlohe, Germany.
6. The western sun casts a beautiful silhouette of the town onto the canal. The large modern building on the left is an unfortunate intrusion on the scene. Ghent. Flanders, Belgium.
7. The line of trees along the canal hides the houses, lending the scene a rural atmosphere. But the shops hidden at the back are fairly showy. The town of Sluis is just three kilometers from the Belgium border. The Netherlands.
8. This town was built at the mouth of a canal which was cut through a mountain limestone rock. Here it is seen from across the canal. Omiš. Dalmatia, former Yugoslavia.
9. The city along the great canal, as viewed from Rialto Bridge. Venice, Italy.
10. The town of Peebles, as seen from the opposite side of the Tweed River. Border, Scotland.
11. The old part of Fribourg, along the river Sarine. Switzerland.

12. 在德克河對岸的遊覽勝地，由道維爾望道而見的爾維爾。　（法國，諾曼第）

13. 沿著古老碼頭的翁弗勒勒街道的房舍。　　（法國，諾曼第）

14. 碼頭邊的房屋及住家。麥伯季聖（英國，康渥爾）

15. 籠罩在暮色裡的蒙特勒城燈光遠映於日內瓦湖上。（瑞士）

12. Trouville, on the Touques River, as viewed from the resort town of Deauville. Normandy, France.

13. Houses of the old part of Honflerur, by the Old Harbour. Normandy, France.

14. Houses on the harbour. Mevagissey. Cornwall, Great Britain.

15. Streetlights of Montreux reflected upon Lake L'eman at twilight.　Switzerland.

# 陸連島

Small offshore islands

　細長的沙洲和陸地相連，就像日本有名的〝江之島〞般的小島，是十分難得的自然景觀，但光是這樣，並沒有到那麼引人注意的程度。法國的蒙・尙・米歇爾(Mont St. Michel)、西班牙的班尼斯柯拉(Peniscola)、舊南斯拉夫(former Yugoslavia)斯本第・史蒂芬(Sveti Stefan)的等陸連島，因島上重新建設，增蓋建築之後，才引起了人們的矚目，進而成爲有名的觀光地。

Spits of connected islands are rather rare in nature, but as such not especially attractive to the human eye. What makes such places as France's Mont St. Michel, Spain's Peniscola or former Yugoslavia's Sveti Stefan such famous tourist attractions is undoubtedly the buildings which stand upon these islands.

1. 曾經是漁村的斯本第・史蒂芬。現在全島皆是高級的旅館了。（舊南斯拉夫，蒙第涅格洛）

2. 被壯麗的城牆所包圍的普特瓦城。它位於距斯本第・史蒂芬不遠的半島之上，其佈景等景觀並不是那麼地好，所以前往的遊客群的水準亦不甚好吧！大多數都是旅行社固定遊程的遊客。（舊南斯拉夫，蒙地內哥羅（黑山）。

3. 班尼斯柯拉。照片中可以看見的城牆，就是中世有兩位教皇的分裂時代，其中一位對立的教皇貝來狄八世的避難處。島上城內街道略爲龐雜，雖也不能稱爲景觀優美，可是仍有大量的遊客至此，所以一定因外圍景觀而吸引觀光客前來的。（西班牙，雷本特）

4. 蒙・尙・米歇爾則是因爲其特異的景觀之故，加上古代世界七大不可思議之故，被譽爲〝西歐的不可思議〞。（法國，諾曼第）

1. Once a fishing town, the entire island of Sveti Stefan is now a luxury hotel. Montenegro, former Yugoslavia.
2. The resort town of Budva, surrounded by its beautiful castle wall. Though not far from the island of Sveti Stefan, Budva is actually on a peninsula, and therefore suffers somewhat as a setting. Perhaps for the same reason, most of the guests seem to be on package tours. Montenegro, former Yugoslavia.
3. Peniscola. Pope Benedict the Eighth fled to this castle during the Great Schism, in medieval times. The town within the walls is a mixture of various architectures, and far from lovely. The large number of tourists are no doubt attracted by the setting itself. Levant, Spain.
4. The Mont-St-Michel, because of its incredible setting, is added to the Seven Wonders of the Ancient World as the "Wonder of the Western World." Normandy, France.

# 崖

Cliffs

　岩壁，斜面的山岩，懸崖，以自然景觀角度而言，就是十分引人注目的，更不用說，建在山崖上的村落，以山崖爲背景的村鎮，還有建在懸崖上的村鎮，更是受到大家極度的矚目。削尖挺立的大崖石本身就令人感到畏懼及壯觀，看到那些建在山壁上的住家時，畏懼之情立刻被驚訝和讚嘆所取代。因此，那些傍崖而建的村鎮，給人的印象更是深深烙印在腦海中。

Rock escarpments, walls and cliffs have always been elements of nature which attract interest, but when villages and towns are set before them, or above them, or even hanging against them, then they become doubly impressibe to the eye. Sheer precipices have the power to instill fear in people, but when dwellings are seen upon them, the fear becomes appreciation and rejoicing. The image of a village or small town against the backdrop of a cliff leaves a powerful and permanent impression on the beholder.

2.　建在幾乎垂直的崖壁下，充滿不安感覺的村落，古露格（法國，卡露西）

1.　因多爾多涅河的支流阿魯茲河的慢蝕所造成的峽谷斜面上，建有中世巡禮的城堡，羅卡馬度魯。這個中世紀的城堡是建在距河床125m的高崖上，而下方的絕壁上則有以前的民宅、教會、尖塔和拱門等，錯雜地重疊著。（法國，卡露西）

3. 照片1另一角度羅卡馬度魯的景緻

4. 築在可以俯視多爾多涅河的崖上的貝納格城和可以沿著陡徑前往的各住家及村落。（法國，貝利格爾）

5. 建在阿爾卑斯U型山麓的拉烏達布魯爾村和蘇達烏布帕瀑布。

6. 以陡峻的石灰岩山壁爲背景構築而成的村莊。姆斯第耶・沙德・馬林 （ 法國・普羅旺斯 ）

1. Incursions of the Alzou River, a branch of the Dordogne, created a gorge upon which the medieval holy site of Rocamadour was placed. A castle stands atop the cliff, 125 meters above the river. Beneath it is a complex of houses, chapels, arches and gates, all cling to the sheer rock face. Quercy, France.

2. The somewhat unsettling view of Gluges, built beneath a nearly vertical face. Quercy, France.

3. Rocamadour as seen from the opposite side as photo 1.

4. Beynac castle on the cliff above the Dordogne River, and the village of houses built along the steep footpath leading up to it. Perigord, France.

5. The village of Lauterbrunnen in its U-shaped valley in the Alps, and Staubbach waterfall. Switzerland.

6. A village built against the backdrop of a sheer limestone mountain. Moustier-Ste-Marie, Provence, France.

## 狹長土地及斜坡
Narrow lands and slopes

　　雖然不像崖壁般會給人深刻難以磨滅的印象，
但築在狹長土地及斜坡上的村鎮，亦有其獨特吸
引人的魅力。魅力所在有二：其一爲這種景觀下
的村落是無法造成複雜、擁擠的高密度住宅的；
其二是因村落皆建在斜坡上，是可以看見其重疊
交錯的立體構築。另外，這種狹長土地及斜坡上
的村落，大多爲漁村或爲港口城鎮。若是良好的
港口，即使是在狹長的土地上也會有很多人口聚
集於其上。因大家的房屋都是建在斜坡之上，所
以視野十分良好。一般的住宅或是別墅，爲求良
好的眺望景觀，通常都將房子建在斜坡之上，這
樣的建築物群集於此而形成了這幅饒富趣味的別
緻風貌。

Villages and towns situated on narrow lands and slopes
are not as spectacular as those found upon cliffs, but they
are eye-catching. One of the reasons is that houses in
these settings are often crowded tightly together, creating
a complex pattern. Another reason is that the layering of
structures results in a three-dimensional view. Fishing
villages and harbor towns are often built in such places,
and are often quite crowded with residences.

　　An advantage of building on a slope is that the views
are good. Originally, houses and villas may be built in
such places simply for the good vantage, but as they
accumulate, they can add to the fine scenery.

1.　台地下海岸部份的狹長平地上，是漁家密集聚居之處。班拿（蘇格蘭）

2.　德國式尖型屋頂的型房屋並排於斜面之上。阿
　　爾坦遜泰克（德國，黑森林地區）

1. Fishermen's homes crowd a slender strip of
   coastline beneath the plain. Penan, Scotland.
2. Classical German-style gable roof homes rise
   one above the other on the slope. Altensteig,
   the Black Forest, Germany.

3. 白色盒子般的房屋彷彿訴說著受伊斯蘭文化影響地廣布於面海的綠色斜坡上。些避署小屋原來都是當地漁夫的住家。保基塔諾（義大利，阿馬爾菲海岸）

6.

4. 阿馬爾菲海岸的中心，阿爾馬菲密集地充滿了西班牙風的高聳白色建築物。在十至十一世紀時，這個地方是和威尼斯並列爲國家的兩大海運中心。（義大利）

5.

5.6. 西班牙北部面對大西洋的小漁村，古提季洛。在港口的斜面，有很多褐色屋頂的房屋緊圍著。爲能活用狹長的土地，所以即使是小村落，也可以看見4～5樓的高層建築。（西班牙，科斯達，威爾第地區）

3. White box-like homes show the influence of Islamic culture on this green slope facing the sea. These resort homes once belonged to fishermen. Positano. Amalfi Coast, Italy.

4. Amalfi, at the center of the Amalfi coast, shows Spanish atmosphere in the shape of its high-rise white buildings. In the 10-11th centuries this was a seafaring capital on par with Venice, Italy.

5, 6. The rustic fishing village of Cudillero sits on the Atlantic coast of northern Spain. The slopes surrounding the harbour are densely covered with houses with brown tiled roofs. To make the best use of the confined space many buildings stand 4-5 stories high. Costa Verde, Spain.

# 山丘上的聚落

Hilltop Villages

在歐洲（特別是南歐），以防衛為目的地將房屋聚落建築在山丘上而殘留情形很多。建在平地的村鎮，可由其外側看見村落內的房屋及高塔等，但整體的印象卻不易獲得。可是，建在山上的村鎮，卻可以由周圍仰視地窺得全貌。另外，山上的村落或城鎮的構築，通常都是緊湊結集的。由山上的村鎮，向外可以很清楚一目了然地自看清村鎮的一切，給人十分深刻的印象。

When seen from flat ground it is hard to determine the shape of a village or small town. One can see houses on the fringe or towers, but it is hard to form an overall impression. With hilltop villages and towns one can look up from all angles, and it is also true that hilltop viliages and towns tend to be compact, so often all that is required is one look from the outside to see the whole village or town. The impression of such a view is lasting.

1. 急在陡峻的山岩上，密集且緊鄰而建的石造房屋。貝伊揚（法國，里維埃拉）·

2. 蒙特福里歐堆在山頂上的教會和塔房，令人印象深刻。（西班牙，安塔露西亞地方）

3. 建築在山岩上的村莊，聖安德尼諾。（法國，科西嘉島）

1. On the crown of this precipitous hill stands Peillon, a cluster of stone houses. Riviera, France.
2. The church and tower at the top of the hill leave a lasting impression. Montefrio. Andalusia. Spain.
3. Sant-Antonio, a village carved into a hilltop. Corsica, France.
4. Gordes, the Acropolis of Provence. France.
5. St. Paul, a hilltop village surrounded by 16th century castle walls. Riviera, France.
6. Èze, a typical hilltop village built to be defensible against attack from the sea. Riviera, France.

4. 被形容為普羅旺斯衛星城的山丘聚落。科爾多
（法國）

5. 被建於十六世紀的圍牆圈住了山丘上的小城鎮，聖·
保羅。（法國，里維埃拉）

6. 為防備由海上入侵者而建於山丘上的典型村落，埃茲。（法國，里維埃拉）

7. 都拉圖·蘇爾·露（法國，里維埃拉）。爲了
防衛，將房舍以連結方式建築，而未建造城
牆。

8. 在黃土之上建築之聚落，露西昂（法國，普羅
旺斯）。村落內之城壁，都是由略帶紅色之黃
土來塗抹的。

9. 蜜色的砂岩建築物，建滿在山丘的聚落上。巴
爾斯（西班牙，科斯達，巴拉伐地區）

7. For defense purposes, the houses of
   Tourette-sur-Loup are grouped to form a wall,
   but are not fortified. Riviera, France.
8. Roussillon (Provence, France), the village built
   atop a hill of ochre earth. The walls of the

town houses are coated with reddish orchre.
9. Pals, a hilltop village of honey-colored
   sandstone houses. Costa Brava, Spain.

# 輪廓線
## Skylines

村落外側的景觀並不只和佈景有關；屋舍的輪廓，特別是輪廓線（Skylines）也是決定外在景觀的要素之一。從前的村落和城鎮中，教會樓塔（或是城堡等）高於一般建築是十分平常的。因此，通常村鎮的輪廓線是以教會（或是城堡等）的高塔為主而形成的。

Not everything in the exterior view of a village or town is setting. The silhouettes of the buildings, in other words the skyline of the town, is a decisive factor in the exterior view of the town. In olden times villages and towns were built around churches, town halls or castles, and these structures tended to dominate the view. Generally speaking, the skyline of a village or town is dominated by a church, town hall or castle towers.

1. 在坡度平緩起伏狀的翠綠田園上，中世紀貴族們相互競爭而築的高塔林立。聖米尼亞諾的輪廓。（義大利，托斯卡納）

2. 建於石灰岩台地著名的〝七塔城〞，馬特爾，便是以其七個塔尖為構成的輪廓線。（法國，卡露西）。

3. 越過陶爾巴川遠眺被中世城牆圍繞的羅膝堡的外觀輪廓。（德國，巴伐利亞地區）

4. 文藝復興時的建築先驅，普魯涅勒斯所建大教堂的圓形巨大屋頂或是吉爾特的塔或是普威奇爾宮殿高94公尺的鐘樓等，構成了文藝復興時期的城鎮，佛羅倫斯輪廓外觀的特徵。（義大利）

1. The skyline of San Gimignano, a town of many towers built by noblemen in medieval times in this agricultural region of rolling hills. Tuscany, Italy.
2. The town of seven towers, Martel, built on this limestone plateau. Quercy, France.
3. The skyline of the walled town of Rothenburg. as seen from across the Tauber River. Bavaria. Germany.
4. A dome of the cathedral built by the leading renaissance architect Brunelleschi, a tower (campanile) designed by Giotto, and the 94 meter high bell tower of the Palazzo Vecchio are distinctive elements of the renaissance skyline of Florence. Italy.

# VILLAGESCAPE &
# TOWNSCAPE
## ●街道之排列

　　街道之排列等，和建築物的形態及設計景觀同
樣地對建築物間之開放空間（也就是廣場及街
道）的大小及形態有關。

　　文藝復興之後的歐洲，開始對街道及廣場進行
以對稱及背景為準，以直線及幾何形狀的都市形
態設計。但在中世紀所造之街道及廣場，實際上
常是屬於不規則形狀。對中世紀狹窄的市街而
言，開放空間是十分重要珍貴。所以彎曲的小
徑，在中世紀是隨手可見的。因從以前開始一點
一滴地改變，演化而成的村莊，建築形態及式
樣，絕不會有相同的。

　　街道及廣場的景觀，並不只是限於開放空間的
大小或形式，還不如說是並排在側的建築物，給
人更深刻的印象。有時不規則狀的街道及廣場，
及決不相同的建築物構成的街景，會讓人對其美
麗興起無限的讚歎。特別是這樣的街景，通常我
們會形容成是宛如身在圖畫中般。

1. 格根巴哈。（德國，黑森林地區）

The appearance of a townscape is related, along with the
architecture and design of its structures, to the size and
form of the open spaces between structures, which is to
say its streets and squares.

　　Since the renaissance, European streets and squares
have been designed along the principles of symmetry and
perspective, and thus are linear and/or geometrical in
form. On the other hand, streets and squares dating from
the middle ages are frequently of irregular shape and size.
In the middle ages space within towns was at a premium,
and it is not rare to come across narrow, twisting streets.
Moreover, over time towns and villages have been rebuilt,
and new styles of architecture introduced, with the result
that styles and forms are rarely unified.

　　But the views along squares and streets is not so
much a matter of their size and design as it is the
impression made by the structures that line their borders.
There are times when an irregular line of structures along
an unplanned square or street offers a view of exceptional
beauty. Often this type of scene results in what is thought
of as a "picturesque" townscape.

2. 羅滕堡。（德國，巴伐利亞地區）

## 街道的形狀
Street Configurations

　　自古以來，自然形成的村莊聚落，其內部的街道，一定不會是筆直的直線狀。一般而言街道多半是橫越城鎮或是曲折的繞著房舍，更或者是T型，Y型路等，還有些是在路上設置建築物，城門或雕刻等，造成視覺上的效果，實際上，村落城鎮的街道或廣場，我們一定都覺得是位於村鎮中心區的，是因為這些街道（或是廣場）並不會直接地，筆直地通往村外之故。如果這些村鎮的街道和直接延伸到其他的市鎮等，那麼這個村落豈不是和現代的幹道街市一樣，是十分令人感到脆弱且不安。現代的幹道街市是醜陋，不具美感，並且讓人感覺不愉快的。美麗使人心曠神怡的街道，即使是給人寬廣感覺，其街道也不會直接通往村鎮之外，在視覺上，這些使人感覺愉快的街道是屬於封閉式的，所以會該人產生一種包容的安全感。

Streets never pass directly through villages and towns which have developed naturally over long periods of time. Village and town streets turn corners, bend around, and form T and Y junctions. Frequently streets end at buildings, pass through castle gates or by monuments.

　　Actually, the sense of security one feels walking on village streets or a town square has much to do with the layout of the surrounding streets, and the fact that the street do not extend straight away. If villages did have streets that extend straight away, they would appear like the ribbon developments seen along modern highways, which because of their design have a weak, unstable appearance. Ribbon developments are ugly and unpleasant. Towns which are beautiful and pleasant, do not have streets which simply lead away. Visually, villages and towns, even when their streets are wide and spacious, appear at least to some degree enclosed, and this lends them a feeling of cloistering and securiy.

3

4.

3.4. 徒步專用道，其藉緩坡且蜿蜒的小徑，創造出使人的視覺上和心靈上的平靜空間。拉文納（義大利）

1. Gegenbach. the Black Forest. Germany.
2. Rothenburg. Bavaria. Germany.
3. 4. A gently curved street for pedestrians only creates a closed-off effect, which in turn lends a secure ambience to the town. Ravenna. Italy.

5. 連結左右兩側建築物的走廊下方，窄窄的小徑前端彎曲而看不見盡頭。威什薩（義大利，威尼托）。

6. 平緩彎曲的鄉村小徑和旁邊吸引人的小型別墅。吉爾布洛依（法國，皮卡爾第地區）。

7. 西班牙南部白色村舍間的狹窄小徑。阿爾科斯·第·拉·弗倫特拉（西班牙，安塔露西亞）。

8. 對徒步行人造成封閉視覺的羅貝亞（舊街道的購物區）。（法國，柏甘蒂）

5. A street with a corridor bridge which connects the buildings on the right and left sides, and then bends away. Vicenza, Italy.

6. Charming cottages line this gently curving street in Gerberoy. Picardie, France.

7. A narrow street in a white walled village of southern Spain. Arcos de la Frontera. Andalusia, Spain.

8. The visually enclosed shopping area of the old part of Avallon, where the street is for pedestrians only. Burgundy, France.

11. 古留意耶魯的中央大道，正面爲教會，中央大道左邊小徑因彎曲而造成廣場上的視覺封閉效果。（瑞士，弗里堡州）。

9. 羅滕堡的中央街道，從溫特雷·休密特開始街道一分爲二，兩條街道都各有一高塔爲門戶，造成一封閉的視覺效果。
（德國，巴伐利亞州）。

12. 皮布爾斯的中心街道。此街道突然向右彎曲，造成了視覺上的封閉效果。（蘇格蘭，波坦地區）。

13. 皮布爾斯的中心街道12的另一面的反向照片。道路的正面有教堂，仍造成了視覺上的空間封閉感。

10. 沿著威尼斯運河兩岸的街市。這座運河的前端和大運河以T字型相交，所以仍營造出屬於封閉空間的視覺效果。（義大利）。

9. The center of Rothenberg, Untere Schmiedgasse, is separated into two streets, both of which have tower gates which contribute to a visually enclosed effect. Bavaria. Germany.

10. Buildings along a canal in Venice. Because this canal intersects a large canal ahead, the space feels well contained. Italy.

11. The large street in the center of Gruyères. The street narrows and turns to the left in front of the church, achieving an enclosed effect. Fribourg, Switzerland.

12. The center of Peebles. At its middle the street turns slightly to the right, creating a visually enclosed space. Border. Scotland.

13. Peebles as seen from the opposite direction as shown in photo 12. A church blocks the front of the street, once again creating a contained atmosphere

## 建築物的排列方式
### Building Fronts

眺望排列美觀且賞心悅目的街道，可以感受到建築物並排的連續性，而不是房屋零零落落地散布，同時也可以感受到一股單純、愉快的氣氛。

因此，仔細觀察成群的建築物，便不難發現建築物的排列有其整體的協調性，由以下的法則，我們就可以確實地想起這些事實。

1.「屋頂傾斜度」當我們感受到建築物成群美感的同時，也表示我們認同了這群建築物屋頂的統一性及協調性。即使是尖型屋頂、組合屋頂、日式的尖頂四柱型屋頂等，有各式各樣的屋頂型態及各種大小不同的樣式，材料也不一致，但是只有房屋的屋頂傾斜度，卻是幾乎相同的。

2.「窗戶的相稱及比例」當我們感到建築物整體的美感時，即使那些建築物的式樣不一致，即使牆壁的材料不統一，但是其窗口的比例及相稱（窗戶的長寬比例），大致上一定是相同的。並且，窗戶的大小及在牆壁上的比例，也大致是相同的。

3.「色彩的協調」當我們覺察到建築之美時，該建築物整體的色彩也必定是統一協調的。即使各式各樣鮮明的色彩，卻也有可能讓人覺得賞心悅目。這就是我們能認同其色彩協調及其整體性的緣故。

When viewing a beautiful streetfront one notices a simple fact, namely that structures are always grouped together, never scattered here and there. Next, looking more closely still, one sees that in that beautiful group of buildings there is a certain "unified feeling". The following factual observations might also be made:

1. [Roof Pitches] When sensing that a group of buildings is beautiful one will notice that roofs tend to be of similar appearances. For instance, they might be of various types, say gable or hip, or yet again hip-and-valley, and if not of the same materials then at least all without exception pitched at similar angles.

2. [Proportions of Windows] When sensing that a group of buildings is beautiful one might find that building styles do not match, or building materials vary, but that the proportions of windows are almost perfectly alike in each structure.

3. [Coloring] When sensing that a group of buildings is beautiful it is still possible to find that, overall, the coloring of the buildings is not unified. But even if the buildings are variously colored, or all very showy, a good overall blend of coloring will also create a beautiful impression.

1. 完全符合「屋頂傾斜度」、「窗戶的相稱及比例」、「色彩的協調」三大原則的例子。雖是稍稍鮮豔的色彩，但是為市街上帶來明朗的氣氛。傑德巴拉（蘇格蘭，波坦地區）。

2. 屋頂傾斜度大致相同的類型屋頂，左右相連的百葉扇鐵窗，色彩稍有脫落的蠟筆色系，營造出住宅品質高尚的環境。頂克魯斯畢爾（德國，巴伐利亞州）。

1. An example where the pitch of roofs, the proportions of windows, and the blend of coloration — the three basic elements of urban design — have been carefully preserved. The fairly brilliant coloring gives the town a bright atmosphere. Border, Scotland.

2. Here the gable walls, the louvered shutters opened on the both sides of windows, and the fairly low-key pastel coloring gives the houses a high quality appearance. Dinkelsbühl, Bavaria, Germany.

3. Half-timbered houses of Stratford-upon-Avon. If one looks carefully, it is apparent that the windows of the houses on the sides are of medieval shape, while those of the center house are typical English vertical sash windows. But because the window frames are all painted black, the uniformity of the scene is not broken. Warwickshire, England.

3. 艾馮河畔斯特拉特福的半木材式房屋。注意一看，會發現兩側房屋的窗戶都是中世紀的特有形狀，只有中央的房子，上下均是一般英國式滑輪拉窗。但是，因爲在木製的窗檻上塗了強調式的黑色線條，所以房屋的整體協調性完全沒有受到影響。（英國，沃里克州）

4. 由斜面相連的畫一性小住宅群。只不過是一間間廉價的小住宅，但其有趣的連結方式，使得街景也跟著生動有趣起來。如果這些房子不是以如此有趣的方式相連結，而是獨立建築各不相連時，則就會因其廉價而感覺悲慘，可憐吧！蒙特聖安傑洛（義大利，阿普利亞。

5. 雖然屋頂和牆壁的材料各異，但因其吻合「屋頂傾斜度」、「窗戶的相稱及比例」、「色彩的協調」的三大原則，所以街道間反而產生了統一性。哈靈頓（蘇格蘭，東羅吉安）。

6. 建築物的高度不一，壁面上的色彩及材料也各有不同，但是在「窗戶的相稱及比例」這點上有了統一性，因此適度地調整其整體的協調，提高了街道本身的魅力。翁弗勒（法國，諾曼第地區）。

7. 坡道上的鄉間小屋。間或混有茅草屋頂，壁面的顏色及製造方法亦不同，但其最大的魅力則是來自於其符合了「屋頂傾斜度」、「窗戶的相稱及比例」、「色彩的協調」的三大原則。（英國，多塞特）。

8. 克爾蒙舊街市中漂亮的街景。（法國，阿爾薩斯地方）

4. Lined up along the slope is this perfectly uniform group of homes. The homes are nothing-more than inexpensive cottages, but they have own appeal. Set apart, these structures would impart nothing but a miserable feeling. Monte Sant'Angelo.
5. Here again roofing and wall materials are different, but the three key elements — pitch of roof, proportions of windows, and blend of colors — are maintained, giving the town a unified appearance. Haddington. East Lothian, Scotland.
6. Buildings of varying height and color, but with windows of equal proportions, can lend a townscape a suitable degree of variety, and even increase its appeal. Honfleur. Normandy, France.
7. Cottages on Shaftesbury's famous Gold Hill. This mixture of thatched roofs, varying walls and different colors makes this a truly attractive street. The balance of the three key elements — pitch of roof, proportions of windows, and blend of colors — is clearly at work. Dorset, England.
8. A beautiful street front in the old section of Colmar. Alsace, France.

# 色彩
Color

對各建築物而言，其色彩上的協調是極其重要的。但是比建築物本身的色彩更重要的是，當建築物並排成列時，和其他建築物間的色彩調和度。如果建築物間有一特定的基本色系，則配合此基本色系建築出來的，就是街道色彩統一，給人時髦整體感的村鎮。如果，各建築物本身各有其鮮麗色彩，則就可能營造出生氣勃勃，明朗氣氛的聚落。可見得街道間的色彩是決定該村鎮魅力的要素之一。

Color is a very important aspect of architecture. But even more important is the coordination of coloring among buildings which stand together. When coloring of buildings is based upon a single tone or color, the overall effect of a town streetfront can be very chic. On the other hand, where colors are distinctive from building to building, a colorful, festive atmosphere is achieved. The coloring of a town can do much to increase its attraction and appeal.

1. 建築物本身的色彩，構造材質及其上色的方法是十分重要的。梅布爾基（英國，康渥爾）

2. 在房屋外塗上強烈鮮豔色彩的布拉諾島街道。如果只是塗上一般看起來十分高尚的顏色，相信這條街道看起來會是十分枯躁無趣的。（義大利，威尼斯）

3. 街道上全體塗成土黃色的露西昂停車場邊的建築物（請參考照片9）。可能是以油漆塗抹上，藉以達到統一色彩的目的吧！（普羅旺斯地區）。

4. 在坎薩爾貝爾重新粉刷過的半木造式房屋。意想不到的是鮮豔的色彩和這古舊的街道氣氛，可以達成協調和諧性。（法國，阿爾薩斯地方）。

5. 愛爾蘭的街道特徵是，商店前的遮陽棚，都塗上十分鮮豔的色彩。在這裡一棟建築物使用一種顏色。克里夫登（愛爾蘭）。

6. 以前曾是漁村的波爾多菲諾現在已經是超高級的避暑勝地了。這個村鎮的魅力在於綻藍的水面映著一片碧綠，兩旁還有義大利色彩濃厚的街道。（義大利，利古利亞）

7. 半木造式的木製窗櫺上塗上天空色的民宅。吉爾布洛伊（法國，皮卡爾第）

8. 富於色彩的街道和周圍的一片翠綠，遠映在靜謐的水面上。

9. 以黃土塗抹的露西昂民宅。（法國，普羅旺斯地區）。

1. The color and texture of architecture, including the way homes are painted, is extremely important. Mevagissey. Cornwall. Britain.

2. The very showily painted homes of the island of Burano. A typical sophisticated blend of paints would have created a boring townscape. Venice, Italy.

3. A building by the parking lot of the uniformly reddish town of Roussillon (see photo 9). The color of the paint is in keeping with the theme of the town. Provence, France.

4. The newly-painted half-timbered homes of the town of Keyserberg. The showy paint works surprisingly well on an old townscape. Alsace, France.

5. The shopfronts of Irish towns are characteristically very brightly colored. Here each shop uses a different color of paint. Clifden, Ireland.

6. Portofino, once a fishing village and now a luxury resort. The charm of this town is found in the reflection of deep green and distinctive Italian colors upon the cobalt blue inlet. Liguria, Italy.

7. Traditional half timber homes with brightly painted frames. Gerbroy. Picardie, France.

8. A colorful town and surrounding greenery reflected upon a quiet harbour. Tobermory. Mull Island, Scotland.

9. Roussillon's traditional homes painted with the ochre of the region. Provence, France.

# 廣場

The Squares

　　歐洲的村落城鎮設計要素之中，最重要的就是廣場。以教堂和政府機構等重要建築物爲首，和其並列的通常是廣場，因此所謂廣場的設計，其實應該是對廣場周圍築物的設計。另外，廣場上也會有噴泉，或是雕刻等裝飾。所以廣場也可以說是村鎮築空間的中心，同時廣場也是舉辦祭典等活動的中心。

One of the particularly important design elements of European villages and small towns is the square. Such impressive structures as the church and town hall are usually found along the square. Thus, when speaking of a square's design, the buildings surrounding it must also be considered. There are also various adornments found on squares, such as fountains and sculptures. The square is the central architectural space of villages and towns, and likewise the place where important activities such as festivals and fairs are held.

2.　吉馬良斯的廣場，拉魯格‧德‧托烏拉魯。廣場的地上舖著波紋狀的馬賽克嵌磚，花壇上綻放著五顏六色的花朵。在廣場的周圍有一些歷史性的建築物，因此這個廣場可以說是古典型都市設計最好的例子。
（葡萄牙，科爾塔，貝魯德）。

1.

3.

1.3.　錫耶納的坎波廣場。這個廣場中央的扇形部份是以紅磚舖設而成的，周圍的部份是用石頭圍舖起來的。以八條放射線狀地將廣場分成九個扇形，有略爲緩和的傾斜，而在廣場最低的位置，是市府的辦公處，且有一高88m的塔。

1, 3. The Piazza del Campo square in Siena. The central fan-shaped portion of the square is brick, and the area around it paving stones. The eight lines radiating from the center separate the fan-shaped square into nine sections. At the base of the gently sloping square stands the 88 meter Torre del Mangia (tower) of the Palazzo Pubblico (town hall), Italy.

2. The square of Guimarães, Largo do Toural. The wave mosaic paving, colorful flowers and historical surrounding the square make this a splendid example of classical town design.

Costa Verde, Portugal.

4. The Piazza dei Signori of Verona, with its beautiful surrounding buildings. Italy.
5. Ravenna, Italy.
6. Perouges, Burgundy, France.
7. The interesting scale pattern paving stones of the square of Atrani. Amalfi Coast, Italy.
8. The square of the town of Évora, the Placa do Giraldo, Alentejo, Portugal.
9. Batea, Aragon, Spain.
10. Caminha, Portugal.
11. Bouvron-en-Auge. Normandy, France.

4. 周圍環繞著壯麗建築的維羅納的廣場。（義大利）

5. 拉文那（義大利）

6. 佩露吉（法國，柏甘蒂）

7. 亞德拉尼廣場的地面上是魚鱗狀的鋪石，十分有趣。（義大利，阿馬爾菲海岸）

8. 埃佛拉的廣場。（葡萄牙，阿連特如地區）

9. 巴特亞（西班牙，阿拉貢州）

10. 卡米尼亞（葡萄牙，阿拉根州）

11. 布渥隆·安·奧久。（法國，諾曼第地區）

## 階梯及窄徑

　　爲中世紀發源地的歐洲（特別是南歐）的村落及城鎮，稍具歷史地區的街道，往往都是十分狹小的。在汽車廣爲使用前的漫長時期，人們生活上的動作，幾乎完全是徒步進行的，以家畜貨運是非常普遍的事，在那樣的時代裡，窄徑並不會對生活造成任何的不便。因此，在很多地方，即使一開始時是寬廣的大道，但經過時代變遷，道路邊屋舍的重建等，以致成爲現在的狹徑。此外有些山上的聚落等，街道是建在斜坡上，因而傾斜的坡度，自然地就成爲階梯了。雖然現在使用汽車時很不方便，但若是徒步時，則仍是恰到好處的。所以對市區街的變化而言，也可謂是件有趣的事。

Villages and towns which originated in the middle ages, particularly those of Southern Europe, tend to have very narrow streets. Through the many centuries before the appearance of the automobile, most people traveled on foot. Pack animals were used for transport. For this even where roads were originally wide, as buildings were rebuilt they encroached on the street, gradually making for narrower roads. In towns build upon hills, narrow roads on steep streets frequently become stepped. This is unfortunate for drivers, but very nice for pedestrians, and it also makes for an interesting town.

1．佩魯賈。（義大利，翁布里亞）　　2．阿爾柯斯・第・拉・弗隆特拉（義大利，安塔露西亞地區）

3．威尼斯。（義大利）　　4．烏爾比諾。（義大利，馬爾契斯）　　5．貝拉吉歐。（義大利，倫巴第）

6. 阿特拉尼。（義大利，坎帕尼亞）

7. 卡拉聖伊提。（西班牙，阿拉貢州）

8. 蒙特聖安傑洛。（義大利，阿普利亞）

9. 貝伊揚（法國，里維埃拉）

10. 科爾多（法國，普羅旺斯）

11. 塞普利亞。（西班牙，安塔露西亞地區）

# 拱門及拱頂走廊
Arches and Arcades

　　在歐洲的村鎮中，拱門或拱頂走廊是十分常見的。對石造及磚造的建築來說，拱門是爲了大型開口而建的合理或者說是必然的形態。對建築在城壁間村落及城鎮而言，城壁的門一定是拱門，無一例外；而構築跨越通道的建築物時，其構成的部份也必定是拱門。另外橫越街道的建築物，也一定是以拱壁來支撐連結的，這時的拱壁，也是拱門狀的。拱門除了是象徵著該處有街道的標誌之外，同時也並存著美化景觀的作用。在建築物下方連結著拱門的拱頂走廊，在夏天可蔽日，在雨天能躲雨，所以除了其形態美觀之外，同時也有其實際的功用。並且拱門兼具有重覆性的韻律感。特別是在南歐村鎮，面對著主要廣場和街道的建築，大多數都有一層拱頂走廊。

In European villages and small towns one frequently comes upon arches and arcades. In situations where stone (or brick) is the material, arches may be the most rational (or only) way to create a large opening in a wall. Where towns are surrounded by castle walls, arches are invariably the solution to castle gateways. Where buildings are part of the wall, arches are used to create roads, and where narrow streets are surrounded by buildings, once again the buttresses of buildings are created with arches. Arches are symbolic of the presence of a street, but are also beautiful in their own right, acting as a frame for picturesque views. Arcades, which are created by a series of arches, not only shade from the sun and protect from the rain, but have their own esthetic value as well. The repetition of arches has a rhythmical effect. In Southern Europe one often sees arcades formed of arches along the ground floors of buildings.

1. 阿西西。（義大利，翁布利亞）

2. 佩魯賈。（義大利，翁布利亞）

3. 埃久·莫魯特。（法國，普羅旺斯）

4. 傑魯布洛伊。（法國，皮卡爾第地區）

5. 拉文納。（義大利，艾米利亞－羅馬涅）

8. 波洛尼亞。（義大利，艾米利亞－羅馬涅）

6. 巴特亞（西班牙，阿拉貢州）

9. 埃佛拉。（葡萄牙，阿連特如）

7. 曼圖亞。（義大利，倫巴第地區）

10. 歐魯那茲。（法國，法蘭西·康德）

1. Assisi. Umbria, Italy.
2. Perugia. Umbria, Italy.
3. Aigue Morte. Provence, France.
4. Gerbroy. Picardie, France.
5. Ravenna. Emilla-Romagna, Italy.
6. Batea. Aragon, Spain.
7. Mantua. Lombardy, Italy.
8. Bologna. Emilla-Romagna, Italy.
9. Evora. Alentejo, Portugal.
10. Ornans. Franche-Comte, France.

# BRIDGES

●橋

在歐洲面對著河川的村鎮，常常會發現一些美麗令人印象深刻的橋。在面對著河川的村鎮，橋是渡河的主要交通方法，從以前開始，橋就在經濟（或者是防衛）上占有重要的地位，而橋的重要性即表現在橋的設計上。一座設計得宜的橋，在其下的水面上，村鎮的景觀往往可以如同一幅名畫般地呈現出來。在中世紀時，橋樑之搭建通常是由敎會或熱心公益者所爲。此外，在重要地方架設的橋樑，有爲了維修或是領土的收入，加以課程或是收通行費的情形也不少。具有這樣功能的橋樑在村鎮形態下，而發展成交通要衝。

在開始工業革命，運河和鐵路發展完備之前，龐大的貨物最常見的就是由家畜來搭背；自羅馬時期以來，有些橋樑因僅爲家畜通行而寬度狹窄，故被稱爲〝馬橋〞。其中，有些橋墩上會有三角形的突出，這是爲了在狹窄的橋中相遇時，可以擦身而過的權宜之計。到了交通量增加了的現今，馬車成一般的交通工具，橋樑也因應需要而拓寬，但在村鎮之中，仍留有過去〝馬橋〞的踪影。

歐洲的橋樑，是源自羅馬時代以來的傳統，多爲石造或磚造的拱橋。文藝復興之後，橋樑的設計漸漸地轉向科學，著重以構造力學爲設計的基礎。在十八世紀時，鐵橋登場了，以至於二十世紀的鋼筋混凝土橋的發明；雖是如此，但是石造拱橋的基本形態卻是至今沒有改變過。即使石造或磚造橋樑經過長時間而有了些許的變革，但並沒有產生國家或地區的共通的特徵。當然各地會有不同的風格，也會爲景觀帶來各種獨特的色彩。

By European towns and villages which are set along rivers one frequently sees beautiful bridges. These impressionistic structures have been used since ancient times, either for economic or defensive purposes, to facilitate overland traffic, and are thus extremely important. Their significance is obvious in the variety of their designs. The beauty of a bridge, reflected upon the river beneath it, can elevate the setting of a town to the level of a watercolor painting. In the middle ages bridges were built by the churches and lords of the land. Tolls and taxes were collected at tollhouses along most bridges, either to pay maintenance or to pay for the occupation of the country. Towns by such bridges often developed as a result of the traffic.

Until the industrial revolution, when trains and canals came into use as means of transporting goods, pack animals were the chief means of supply, and from Roman times many bridges were built only wide enough to accommodate horse. These were referred to as "pack horse" bridges, and some had triangular protrusions over the supports so that animals could pass each other. When wagons came into common use many bridges were expanded, but today there are still many pack horse bridges to be seen.

In Europe, most bridges are built of stone or brick, with arches, in the tradition of the Romans. After the renaissance, more scientific approaches were used in the construction of bridges. However, until the development of iron, at the end of the 18th century, and reinforced concrete in the 20th, all bridges were built in the basic way, upon arches. Although there are many variations of arched bridge, there is nothing unique about the arched bridges of a particular country. Each and every bridge seems to be distinct, and one therefore finds that villages and towns with bridges all have their own particular character.

1. 位於舊南斯拉夫，黑塞哥維那的中心都市莫斯塔爾的優美拱橋。這座橋橫跨奈雷特瓦河，是座單一拱狀的大石橋，是十六世紀奧圖曼帝國的工程師所建造的。

2. 和照片1同是莫斯塔爾的拱橋,此幅照片爲全景和街道。曾經是如此美麗的城鎮,而現在據說橋已崩壞,不復存在了。

4.

3.4. 這座中世紀的橋是在1962~1965年間修復完成的,貝沙魯,西班牙。貝沙魯的這座架在村鎮入口處的橋,是利用河川中間的砂洲,並且橋間尚有彎曲之設計。在修復前,這座橋八個拱型橋座最中央的兩個,已經完全崩落了,此外,兩個防禦用的塔樓也崩壞了。關於這座橋的修復,西班牙政府將之視爲一項文化財產保護,並且以羅馬的要塞式建築爲藍底復建之。

3.

1. 2. Mostar, in central Hercegovina of former Yugoslavia, boasts this lovely arched bridge. The single arched structure, which spans the Neretva river, and was built by the Ottoman Turkish engineers in the 16th century.
3. 4. This medieval bridge in Besalu, Spain, was rebuilt between 1962-65. The bridge stands at the entrance to the town, and makes use of islands in the middle of the river, where it also makes a bend. The bridge has eight arches, and before repairs were made the two in the middle had completely collapsed, as had the towers guarding the ends of the bridge. As a result of the national culture preservation effort the bridge was beautifully restored to its original Roman fort form. Besalu, Spain.

5. 阿馬蘭特的拱型橋。（葡萄牙）

6. 巴塞盧斯。（葡萄牙）

7. 弗德。（英國，蘇格蘭）

10. 夏爾特爾。（法國）

8. 隆達的村鎮，1971年所建連結南北之拱橋。（西班牙）

11. 中世紀橋成爲往來要塞，而以此聞名的位於法國，卡奧爾的龐特‧瓦朗德雷橋。

9. 弗里堡州的伽利根橋和聳立於崖壁上之街道。（瑞士）

5. Arched bridge in Amarante. Portugal.

6. Barcelos, Portugal.

7. Ford, Scotland, Great Britain.

8. The arch bridge, which crosses the town from south to north, was built in 1791. Ronda, Spain.

9. Zahringen Bridge, and above it the impressive view of Fribourg, Switzerland.

10. Chartres, France.

11. Pont Valentré Bridge in Cahor, which is famous as a medieval defended bridge. France.

12. 著名的利亞祿德橋，座落於威尼斯的大運河，建於1588～1592年。
這座橋是由安東尼奧‧德‧龐特所設計的，拱橋的兩側各用6000枝的木樁打入，上面再藉拱型之推力而安裝直角狀的基石。因為以這種力學為基礎的設計，所以即使是建在柔軟的堆積土，且兩側有商家的大石橋，也能屹立至今而不崩壞，表示在這種環境狀況下，要建出牢固的建築物仍是可能的。（義大利）。

13. 威尼斯。（義大利）

14. 1567～1569年建於著名的佛羅倫斯的特里尼塔橋。這座橋的設計者是巴魯特洛米歐‧阿馬納帝，他應用科學原理來設計，將拱型高度與跨度設計成7：1。這個形狀也被稱之為提藍的把手，在這之後，這種形狀的設計便開始普及了。因此，在歐洲的橋樑之中，凡是拱型呈此種形狀者，表示這座橋的建築年代不如我們所想的那麼久遠。這座橋曾在第二次世界大戰時遭受破壞，戰後則立刻將掉落於河川內的原材料拾回，忠實地復原，重建，使其恢復原貌。（義大利）。

12. The famous Rialto Bridge, built between 1588 92 over the Grand Canal of Venice. The architect, Antonio da Ponte, imbedded 6,000 Timber piles under each abutment of the canal, then laid stones upon them, perpendicular to the arches above. The result of this mechanical design was a stone bridge which was able to support a street of shops, and which still stands today. Venice, Italy.

13. Venice, Italy.

14. The famous Trinita Bridge, built in Florence between 1567 69. The architect, Bartolommeo Ammanati, used mathematical principles to determine an arch height to span length ratio of 7 to 1. This design, termed ''basket-handled,'' later came into wide use throughout Europe, so that one can usually determine that a bridge with long horizontal arches is not ancient. This bridge was destroyed in World War Two, but the materials were later retrieved from the river bottom, and the bridge authentically restored. Florence, Italy.

15. Verona, Italy.

16. Castelneaud, France.

17. The stone bridge at Calw, Germany. The bridge has crenellation on both sides. Calw, Germany.

18. Dordogne, France.

19. Chartres, France.

20. Noyer-sur-Serin. Burgundy, France.

21. Near Vezley. Burgundy, France.

22. St. Crois en Jarez, France.

15. 維羅納（義大利）

16. 卡斯特魯諾。（法國）

19. 夏爾特爾。（法國）

20. 諾瓦耶・斯慧露・斯蘭。（法國，柏甘蒂）

21. 維祖爾近郊。（法國，柏甘蒂）

17. 建於十五～十六世紀的石橋。卡魯夫，德國。在橋的側面，有預留了守備用的鎗眼。

22. 聖・克洛瓦・安・嘉雷。（法國）

18. 多爾多涅地區。（法國）

23. 布魯日。（比利時）

25. 廷克爾斯軍爾的近郊。（德國，巴伐利亞地區）

26. 費爾伯克。（德國）

24. 施韋比施哈爾。（德國）

27. 奧克斯福德。（義大利）

28. 皮布爾斯。（英國，蘇格蘭，波坦地區）

29. 於1970年完成的福斯鐵路橋樑和並行的吊橋福斯道路橋樑。昆士菲利。（蘇格蘭，愛丁堡）

30. 因利用鋼鐵來造橋，所以橋樑的設計也有了大幅的變化，在1890年，佛渥斯爵士建造了第一座懸樑式橋樑，就是福斯道路橋樑，在當時這座橋，超過七年前紐約的布魯克林大橋的跨度，
創下世界最長的跨度（1,710英尺）的紀錄。當時的藝術家威廉·莫里斯曾形容這座橋是極其醜陋的，但是現在這座橋卻受到很多人的青睞。昆士菲利。（蘇格蘭，愛丁堡北部）。

33. 皮布爾斯。（英國）

34. 西班牙和葡萄牙國境橋

31. 皮布爾斯。（英國，蘇格蘭，波坦地區）

# ARCHITECTURE

● 建築

　　在小村鎮裡的建築物，大部份都是我們稱之爲
〝地方建築〞的淳樸民宅。教會、修道院、政府
機關及地主宅邸等，通常這類正式的建築（也可
說是較大規模的建築），都是由建築師或是建築
相關的專家來設計，而業主往往也可以反映出在
國際間，具重要社會階層者的力量及關懷。因
此，這類正式性的建築設計，通常較重視美感，
藉以成爲全國性或是國際性的建築典範。相對於
此，〝地方建築〞是地方上一般人民的生活，文
化背景的反映，相較於美感的表達，他們更重視
的是承襲傳統的機能性設計。這兩種建築設計的
不同，可以十分明確地區分；正式性建築的設計
要素是模仿地方性建築而取其中間型，但也因時
代的變遷而逐漸普及。

　　而在地方建築的設計要素則是，使用的建築材
料和以材料而決定施工方法。

Most of the structures found in villages and small towns
are in what is called "vernacular" architecture, the simple
style of local tradition. Large buildings such as churches,
senates, city halls and the residences of the wealthy are
designed by architects and built by masons and other
specialists on the orders of people of high social, national
or international standing, reflecting the powers and
concerns of these people. Their construction is guided by
aesthetic purposes, and their designs tend to be of
national or international character. On the other hand,
vernacular architecture reflects the culture and living
standards of the locale, and exhibits more of a practical
than esthetic purpose, making full use of local resources
and traditional designs. These two types of architecture
can be fairly well distinguished, although vernacular
architecture in later times often shows elements of true
architectural design in a "half-and-half" style.

　　The key design elements of vernacular architenture
are the building materials and the building methoods
developed for these materials.

1. 埃特魯塔。（法國，諾曼第地區）

# 木造房屋
Timber Construction

木材，分爲橡木等闊葉樹和松樹、樅樹等針業樹兩種。闊葉樹較堅硬，而針葉樹爲柔軟。除了地中海地區之外，大部份的法國、英國、德國和荷比盧等國家，都曾經被廣大的闊葉林所覆蓋，因此在這些地區的小村落和城鎮中，仍留有不少以橡木等堅木築成的半木造房屋。所謂的半木造房屋是表示，木材僅使用一半，而以堅硬木材構築起來的軸組之間，則是使用磚瓦，石村；在木材裝組的底部上漆或是以編好的材塞在土壁的底部。

半木造式房屋，就是以石或磚爲基礎上（或是以石、磚等做爲台階），一樓的部份以結實的組軸爲牆壁的框架，在框架支撐第二階支架（或是屋頂的骨架），在地板架上再逐次地架上牆版架

等，以此結構建築上來。因爲這種構築方法，上層的結構會比下層的牆版更爲突出，所以可以簡單地蓋出上層大於下底的建築構造。事實上，這種半木造式房屋的街道上，常可以看見建築物的上層會稍爲突向街道。

There are two groups of trees used for building materials : broadleaf trees, such as the oak, and conifers, such as pines and firs. The wood of broadleaf trees is hard, that of conifers is soft. In most of France, England, Germany, Benelux and other non-Mediterranean areas, hardwood forests once covered most of the land, and in

the villages and small towns of thesecountries there are many half-timber structures built of oak and other hardwoods. The term ''half-timber'' describes a structure built in part with wood, where, for instance, the gaps in an oak frame have been filled in with brick or stone, or where the laths or wattles have been covered with plaster or earth.

In a half-timber structure, stone or brick is used to form the foundation (or the bottom floor). A sturdy wall frame is then constructed to support the floor of the next story (or the roof, if it is a cabin), and another wall frame built upon that floor to support the next, and so on. In this kind of structure, the floor can jet out beyond the wall below, and one can build each of the higher stories slightly larger than the last. In fact, when walking on the streets of half-timber towns one often notices that the upper stories of structures lean out toward the street.

2. 施特拉斯堡。（法國，阿爾薩斯地區）

3. 佩露吉（法國，柏甘蒂）。（瑞士）

4. 覆以魚鱗狀橫木板的牆壁。拉渥塔布魯涅（瑞士）

5. 桑・斯慧露・拉波畢（法國，卡露西地區）上層樓房全都往街道突出，這是木造房屋才會有的情形。照片3.也是如此

6. 諾伊・休魯・斯藍。（法國，柏甘蒂）

1. Etretat, Normandy, France.
2. Strasbourg, Alsace, France.
3. Perouges, Burgundy, France. The ground floor is built of stone.
4. A wall adorned with fish scale wooden tiles. Lauterbrunnen, Switzerland.
5. St-Cirq-Lapopie, Quercy, France. Wooden construction makes it possible for upper stories to lean toward the street. See also photo 3.
6. Noyer-sur-Serein, Burgundy, France.

# 土造、石造和磚造房屋

Earthen, Stone and Brick Construction

在歐洲鄉下地區（特別是地中海地區等雨量較少的地方），即使是現在也還超乎想像地留了不少土造（或是磚造）的房屋。土造的房屋，通常在外側會有漂亮的修飾，像是修飾爲白色等，由外觀是完全看不出土造的痕跡。要確認是否爲土造房屋時，很簡單地只需將小刀插入試試看便可得知，如果小刀可以插入，就是土造房子，而且因爲房子的牆壁都十分厚實，所以不用擔心插出洞來。

歐洲最標準普遍的建築，其實的石造（或是磚造房屋貼上石板）房屋。教會等建築幾乎無一例外的，都是石造的。據說是羅馬風格和哥德式建築的大教堂，將原先的木造天井置換成石造時，琢磨出來而開始的。

石造建築最主要的就是石頭。建築用的大石塊和加工的石材，並不是到處都有，而且古時候重物的運費非常的昂貴。因此，會使用石材的地方性建築，多是能夠就近進行石材加工的地方，也就是說這種石造建築多是在出產良質石材的地方（例如：英國的科茲威德（Cottswald等地）。無法取得良質石材（以及木材）之處，例如像是地中海地區常見的傳統建築，就是以粗石堆積，再上漆，表面以黃土或石灰來修整。這種時候即使稱之爲石造房屋，外觀上和土造及磚造房子是無法判別的。

由石塊堆積而成的屋舍，如果是外觀上不加以修整的地方建築，則觀察石塊種類，形狀和堆積方式等，可以得知造成這種各家特性不同的色彩和構築方式，進而了當地建築上的特色。磚造的屋舍在多樣性上，不如石造房屋；但是在堆建方法及色調搭配上，只要稍加修飾，便不難判讀其年代及地區。

In the rural areas of Europe (especially in the dry areas of the Mediterranean) one comes across many structures of earth and adobe (dun-dried mud bricks) — more, perhaps, than one might expect. Earthen structures are well-formed on the outside, and when white-washed, might not even be externally recognizable as made of such materials. To determine if a structure is made of earth, one need only probe it with a knife. If the knife goes in, it is made of earth. And one need not worry about making a hole through the wall. Earthen walls are very thick.

In Europe, the standard form of construction is stone, or brick overlaid with stone. Almost without exception churches are built of stone. It is safe to say the Romanesque and Gothic cathedrals were born with the discovery of a means to replace wooden ceilings with stone vaults.

The key feature of stone architecture is the stone itself. A stone that is suitable for building, or for crafting into building stone, is not easy to find. Moreover, in the olden days it was very expensive to transport the heavy stone needed for construction. Thus, vernacular architecture with elaborate stone masonry is limited to areas were there is a good source of stone, as there is, for instance, in Cottswald, in England. In areas poor in both wood and stone materials, such as certain regions of the Mediterranean, the vernacular architecture often relies upon construction with rough stones and rubbles, and

1. 史坦頓。 英國，蘇格蘭，東羅吉亞地區

2. 都柏林。（愛爾蘭）

3. 隆達。（西班牙，安塔露西亞）

4. 奧德華。（法國，卡露西）。半木造式房屋的一樓通常是石造

5. 伯頓‧布拉德斯特克。（英國，多塞特郡）

8. 15世紀建於烏爾比諾的磚造房屋，就是拉斐爾的老家。
（義大利，馬爾契斯）。

9. 在具有庇里牛斯山風格的小國家，安道爾境內，興建房
屋時，在磚瓦外側加貼石片，是這個地區自古以來的傳統
建築型態，也是山居國家保存其聚落景觀的方法。

daubed with clay, lime or plaster. Often these structures, in spite of their stone construction, are indistinguishable from earthen abodes.

Where stone masonry is evident in vernacular architecture, the types of stones, the shapes of stones, and the manner in which they have been put together is often their key to the unique color and texture of a given region's architecture. In the case of brick construction, the variety of styles is not as rich as stone construction, but the manner of assembly, the coloring and the decorative elements do show distinctive regional and historical characteristics.

6. 華倫沙‧德‧米紐。（葡萄牙，米紐）。
貼有稱為阿茲烈久斯磁磚的牆壁，是受了伊斯蘭建築的影
響，也是17～18世紀葡萄牙，巴洛可式建築的特徵。

7. 土造，石造或磚造的房子，雖然看起來外觀不是十分漂
亮，但卻十分容易修補。
塞布利亞（西班牙）。

1. Stenton, East Lothian, Scotland Great Britain.
2. Near Dublin, Ireland.
3. Near Rhonda, Andalusia, Spain.
4. Autoire, Quercy, France. Half timber houses frequently have ground floors built with stone or rubble walls.
5. Burton Bradstock, Dorset, England.
6. Valença do Minho. Minho, Portugal. The structure shows walls covered with the local Azulejos tile. The style is 17-18 century baroque, and shows an Islamic influence.
7. Structures built of earth, stone or brick are easily repaired as long as vanity is no obstruction. Sevilla, Spain.
8. The 15th century brick house, birthplace of Raphael, in central Urbino. Marches, Italy.
9. In the small nation of Andorra, in the Pyrenees mountains, new houses are built with brick and covered with traditional stone rubble, thus maintaining unity with the traditional style and exterior appearance of the mountain villages.

# 屋頂
Roofs

　　對地方性傳統建築而言，最能清楚地顯示地方性的就是房子的屋頂。即使在相當廣闊的地區，建築物的屋頂，也就是指屋頂的傾斜度，材質等，大多是相同的，直到最近都還是如此，當然是一般情形而言。地中海地區屋頂，是由一種稱之爲羅馬磚的半圓筒狀的瓦片，由上，下交互地排列構成的。並且房屋的斜度方面，大體而言是稍緩的角度。是因爲這個地區的雨量稍少，且以瓦片排列的構築方式，一超過三十度時，瓦片便較易滑落之故。不過，最近似乎是在瓦片下方舖以毛氈等，以防止剝落；所以在北方屋頂傾斜度較大的地方，地中海式建築方法似乎也可以行得通了，也就是因爲這樣，而造成了風格和地方特色建築的混亂。

　　在石棉瓦的產地，一般是以石棉瓦爲屋頂材料。石棉瓦易於層狀分割，品質良好石棉瓦，是十分輕薄，因此是十分適於建造屋頂的。也因地方不同，除了石棉瓦之外，有些砂岩等具剝離性的石塊或是砂岩，石灰岩等有相當厚度的石塊，也會成爲建築屋頂的材料；正因爲如此，所以屋頂才會這麼具有地方色彩。近年來，茅草及稻草屋頂的房子大幅減少，這是曾是以往最常見的，但現在村落中也幾乎看不到了。稻草屋頂的屋頂傾斜度較大，較陡（40～60°），因此若是看到屋頂斜度較大的古老瓦屋，以往是稻草屋頂的可能性就很高了。稻草屋頂的缺點在於易燃，因此在乾燥地區，是不會看到稻草屋頂的。新的屋頂材料的普及，使得地方性的建築特色因此而改變。例如：一般認爲是英格蘭和蘇格蘭東部低地的建築特色－紅色磁磚瓦片，以前是輸出的羊毛及醃魚船自荷蘭帶回來的進口貨，而這些進口貨逐漸取代了過去的稻草屋頂。

In vernacular architecture, one of the most distinguishing regional characteristics is the roof. This is to say that, until recently, the pitch of roofs, and the materials used in their construction, have been extremely consistent across a broad region.

In the Mediterranean region, the semi-cylindrical "roman tile" is the common roofing material, with its alternating upward and downward facing layering. Moreover, throughout this region the pitch of roofs is gentle. Not only is rain light in this region, but the tiles tend to slip if laid at a sharper angle than 30 degrees. Recently, a layer of felt has been introduced beneath tiles to prevent slippage, with the result that Mediterranean style tile roofs are now being seen farther north, mixing with the vernacular architecture.

Where slate is produced, it is common to see this material used as roofing. Slate splits into layers, and the best quality slate can be split into very thin slices, making it suitable for roof-laying. In other regions, other types of stone which split evenly, such as limestone and sandstone, are used, and sometimes in fairly thick fragments. These materials impart unique flavor to the regional architecture.

Recently, thatch-roofed dwellings have greatly decreased in number, and are rarely seen in small towns and villages, though they were once quite common.

1. 使用紅色的羅馬磁磚的屋頂。正中央的橢圓形廣場，是古羅馬圓形競技場的遺址。盧卡（Lucca）（義大利（Italy）

2. 不規則狀的石棉瓦屋頂聚落。在現代，這樣的傳統建材是相當昂貴的。在右側建築物中的白鐵皮屋頂，應該是即將加蓋石棉瓦的。（義大利，奧斯塔地區（Aosta, Itayl））。

3. 烏比杜斯（Obidos）（葡萄牙（Portugal））以舊羅馬磁磚蓋成的各屋頂，而屋頂上有造形饒富趣味的煙囪

4. 科茲威德地區，以石灰石板爲屋頂蓋的鄉村建築。庫姆
城堡（英國）。

5. 爲使刮風時，白鐵皮屋頂不致飛起，而於屋頂上置放鎮
壓之石塊。白鐵皮屋頂是最便宜的屋頂材料。拉卡拉荷拉
（西班牙，安塔露西亞地區）。

6. 以石片鋪蓋的土耳其式住宅的屋頂。普希特利（舊南斯
拉夫波的尼亞，墨塞哥維那）。

7. 波形瓦的屋頂。史坦頓。（英國，蘇格
蘭）。

8. 稻草屋頂，班拉第的傳統鄉間。（英
國，蘇格蘭）。

9.

10.

Thatched roofs are quite steep (from 40–60 degrees) so where a house has a tile roof of that pitch, it is possible it was originally covered with thatch. The flaw of thatching is that it burns easily, and is not found in dry regions.

New roofing materials can change the look of a region's architecture. For instance, in the eastern lowlands of England and Scotland, the typical red tile roof is in fact an import from Holland, which was returned in the hulls of ships used to export wool and salted herrings. The original roofing material of these regions was thatch.

9.10.11.12. 烏比杜斯形狀有趣的煙囱。
（葡萄牙）。

11.      12.

1. Roofs of traditional red roman tile. The oval area in the center is a mark of a Roman amphitheatre. Lucca, Italy.
2. A hamlet of irregular slate covered roofs. Today, the materials used on these traditional roofs would be quite expensive. Aosta, Italy.
3. Óbidos, Portugal. Oddly shaped chimneys dot the roofs of some of these ancient roman-style tile roofs.
4. Cottages with roofs of cut limestone in the Cottswald region. Castle Comb, Great Britain.
5. Tin-roofs are overlaid with stones to protect against wind. The tin plate roof is the most affordable covering available. Lacalahorra, Andalusia, Spain.
6. The stone covered roof of a Turkish house. Pocitelj, Herzegovenia, former Yugoslavia.
7. Houses with pantiles roof. Stenton, Scotland.
8. A thatched roof at the Bunratty Folk Village, County Clare, Ireland.
9. 10., 11., 12. Unusual chimneys in Óbidos, Portugal.

# 傳統民宅

Folk Architecture

傳統民宅,和民俗藝品等是一樣,是由地方上的人們,以傳統的技術和感性的心去製作出來的。民宅的設計和建造方式,是各地的木匠,石匠等世代相傳地建造出來的地方特色,所以即使經過長時間的傳承,和前代的連繫仍是密不可分的,也因此這種技術精髓在經過時間的變化,技法的改良之後,仍能保存其淳樸的獨特設計,並承接現代的建築風格。所以即使是在現代國際性漸趨淡化之際,各民宅仍幾乎不受影響,也實在不是件不可思議的事。附帶一提的是在法國阿爾薩斯地區的尖頂民宅,和國境另一端的德國境內民宅,幾乎是無法區分的。

民宅分布的狀況和動物的分佈情形,其實是相同的。在廣闊之處,會有幾乎共通相同的大原則,但仍會有細部上些許的不同特徵。並且,極特殊罕見的形態,更是僅限於小村落才可能見得到。例如在西班牙南部,有種稱爲〝圖魯利〞(Trulli)的簡樸的石造圓屋頂住宅。這般珍貴少有的設計,就如稀有動物一樣,是因該地區特殊的條件,才會自古代流傳至今。

2. 基佛德附近。(英國,蘇格蘭)

3. 爲蘇格蘭國家天然財產的普雷斯頓·米勒。(東林頓,英國,蘇格蘭)

1. 庫姆城堡(英國)

4. 在康屋爾地區,流傳著有關惡魔的傳說。在十九世紀初,有一個村人因恐惡魔侵入,爲不讓惡魔有藏身之處,而建了五間圓形,且屋頂豎有十字架的房子。費里安(英國)

5. 布維隆‧安‧奧吉。（法國，諾曼第地區）

8. 巴爾傑姆。（法國，普羅旺斯）

6. 卡斯特魯諾。（法國，卡露西地區）

Like folk art, folk architecture is a product of local traditional technology and feeling. Traditional houses are the work of carpenters and stonemasons who have passed their craft from father to son for generations. Over time revisions have been introduced, resulting in the rustic but unique designs we see today. It is hardly surprising that vernacular architecture is little influenced by recently drawn national boundaries. For instance, the steeply pitched roofs of houses in France's Alsace region cannot be distinguished from those on the German side of the border.

Unlike palaces and castles, traditional homes are invariably built with only materials which can be found locally. The local flavor of towns and villages is thus the direct result of the flavor of local materials.

The distribution of traditional homes is somewhat like the distribution of plants and animals. Over a wide area some formal resemblances can be found, but as one studies more detailed types the distribution narrows. Where features are extremely unusual, one will only find a type in a very confined area. As examples one might point to the cave residences of southern Spain, or the stone-crafted dome houses of Trulli, in Italy. Like rare animals, these rare architecture's are the result of adaptation to local conditions, and have been preserved from ancient times until the present.

7. 盧夫勒。（法國，卡露西地區）

1. Castle Comb, Great Britain.
2. Near Gifford, Scotland.
3. Preston Mill, part of the National Trust of Scotland. East Linton, Scotland.
4. Cornwall has many legends of demons. Early 19th century, they built five circular homes, with crosses over the roofs, offered no corners behind which demons could conceal themselves. Veryan, Great Britain.
5. Beuvron-en-Auge, Normandy, France.
6. Castelneaud, Quercy, France.
7. Loubressac, Quercy, France.
8. Bargéme, Provence, France.

9. 羅滕堡。（德國，巴伐利亞州）

11.

12.

11.12. 黑森林地區的農家。（巴登－符騰堡，德國）。

13. 安塔露西亞地區，隆達附近的農家。（西班牙）

10. 聖特堡。（西班牙，加泰羅尼亞）

14. 瓜地克斯的穴居聚落。（西班牙，安塔露西亞地區）

15. 奧斯塔地區的農家。（義大利）

16. 托斯卡納地區的農家。（義大利）

17. 阿爾貝洛貝洛。（義大利，阿普利亞）

18. 提契馬尼（斯洛伐克）　Cicmany, Slovakia

19. 西第亞洛（斯洛伐克） Zdiar, Slovakia

20. 莫斯塔爾的土耳其人住宅（舊南斯拉夫，波斯尼亞‧赫爾茲格維那 A Turkish house in Mostar, Bosnia-Hercegovina,
former Yugoslavia

21. 布爾日（比利時） Brugge, Belgium

22. 費雷（荷蘭，謝蘭島州）
Veere, Seeland, Holland

## 教會及修道院

對於有領主，騎士或甚至是文盲的中世紀歐洲封建社會而言，各修道院是一群可以讀寫的人們所維繫的高度文化之宗教共同體。其中，位於法國，柏甘蒂（Burgundy, France）的古烈尼（Cluny）的古烈尼教會總部，是個擁有輩出人才的組織，在那個時代的建築形態，後來被稱作是羅馬式建築。修道院時代的羅馬式建築一直持續著，到了各大教堂興建時就成了所謂的哥德式建築。

到了建築有所謂的型式概念時，應該是在文藝復興建築和巴洛克建築之後的十七世紀；在十七世紀之後，以建築型式這個名詞來代表整體的技術及藝術，並窮究其奧妙者，仍屬教會建築。在中世紀時，人們將其收入的十分之一的稅收交給教會，因此教會有這樣的財力作為背景，也就是這個原因，在小村落和城鎮中的教會建築才能異於民宅，成為一有特殊風格，壯觀的建築。

隨著科學的進步，教育的普及，教會的權威也逐漸式微，在現代社會，連星期日上教堂的人都已減少了。話雖如此，但中世紀所留下的寶貴資產之一的教會，仍是小村鎮中不可或缺的重要景觀之一，是人們的信仰和誇耀的象徵之一。

1. A masterpiece of octagonal Byzantine architecture, San Vitale Church was built in 547 by Pope Maximian. The interior of the apse is decorated with a beautiful mosaic. Ravenna, Italy.

2. A small circular church, built in the 5-6th century. Perugia, Italy.

3. The church at the monastery in Clervaux. The style is Romanesque, but the church was actually built in the 19th century. Luxembourg.

2. 建於西元5～6世紀的圓形小教堂。佩魯賈（義大利）

1. 這個有八角形平面設計的拜占庭式建築傑作，聖‧彼得勒教堂是在547年，因麥克西米安教宗而興建的。教堂的正殿內鑲飾著馬賽克圖紋。（義大利）

3. 克雷爾伯修道院的教會。雖是羅馬風格的建築，但卻不是建於羅馬時期，而是建於十九世紀。（盧森堡）

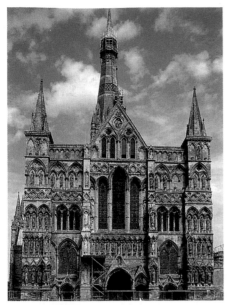

4. 索爾茲伯里大教堂，因具有高聳的左右翼尖塔，而被公認為是英國哥德式的理想大教堂。

5. 具有初期羅馬風格的特里爾大教堂。（德國）

6. 巴洛克建築的傑作，在壯觀的樓梯上，建有新古典主義式的教堂。（龐‧傑杜斯‧德‧蒙特，布拉加近郊，葡萄牙）。

In feudal Europe, where even lords and knights were often illiterate, the religious communities of the Church were islands of enlightenment which maintained a very high culture through the ages. The monastery of the Cluny sect, in Burgundy, France, was an extremely powerful organization which included many popes among its adherents. It established the architecture which was later referred to as "Romanesque." The age of Romanesque monasteries was followed by the gothic period, which produced the great cathedrals of Europe.

Actually, it was after the appearance of the renaissance and baroque archtecture of the 17th century that the concept of an architectural "style" came into existence. But it was upon the different types of early churches that the notion of "arcane architecture," and thus all "styles" of architecture, originated. In the middle ages, people were expected to pay a tithe of one tenth of their income to the church, and it was with this income that structures of distinctive style were built in the midst of traditional towns and villages.

The dissemination of science and learning eventually stole the prestige of the churches, and recently the number of people attending church on Sundays has decreased. Nevertheless, churches continue to be important to the settings of town and villages, and function as symbols and faith and pride to the local people.

7. 位於羅吉諾夫‧波杜‧拉多荷西提姆的野外博物館教會中的木造教堂。（捷克，摩拉維亞地區）

8. 在比薩的羅馬式建築（12～14世紀），聖米凱雷教堂。盧卡（義大利）

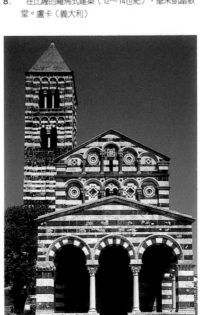

9. 十五世紀由阿爾貝洛第設計的聖瑪麗亞諾瓦教堂。佛羅倫斯（義大利）

10. 修建於十二世紀的天主教派修道院，薩薩拉近郊的聖三位一體教會。（義大利，撒丁島）

4. Salisbury Cathedral, with its great transept spire, is called the ideal cathedral of England.
5. The cathedral of Trier, built in the early Romanesque style. Germany.
6. Bom Jesus do Monte. The neo-classical church was erected at a top of a monumental Baroque staircase. Near Braga, Portugal.
7. The wooden church at Roznov P. Radhostem. Moravia, Czech.
8. San Michel Cathedral, in the Pisan Romanesque style. 12-14th century. Lucca, Spain.
9. Santa Maria Novella Church, with the facade designed by Alberti. Florence, Italy.
10. Holy Trinity Church, outside Sassari, was originally the church of Camaldulian Abbey. Sardinia, Italy.

11.　建於——三六年，幾乎沒有裝飾的西都教派修道院，洛
托羅涅修道院。（法國，普羅旺斯地區）

13.　同樣是西都修道院的迴廊。豐特內修道院。（法國，柏甘蒂地區）

12.　較洛，托羅涅修道院遲約十二年建的另一西都教派的姐
妹修道院，塞奈克修道院。

14.　在西元1100年，克呂尼教會修道院，聖佩特羅教堂的迴廊。莫阿薩克（法國）

11. Le Thoronet Abbey(1136), exhibiting the
    very austere style of the Cistercian Order.
    Provence, France.
12. Twelve years after the construction of Le
    Thoronet, Sénanque Abbey was built as its
    sister abbey.
13. Corridor of another abbey of the same sect.
    Fontenay Abbey, Burgundy, France.
14. The cloister of St Pierre Church, which was
    built around 1100 as the Cluniac Abbey.
    Moissac, France.

17.

15. 巴塔烈修道院最大迴廊西北隅的噴泉。

16.

18.

16.17.18. 巴塔烈修道院,是爲了紀念1385年時和西班牙在
此戰爭並獲勝,故於勝利後的三年開始修築。最大迴廊是
哥德式中,稱爲曼奴埃爾式的葡萄牙式獨特的纖細複雜圖
紋式樣的修築,這是由波伊塔克首創,並完成於十六世紀
的作品。

15. The fountain(lavabo) in the northwest
corner of the Royal Cloister at Batalha
Abbey.

16. 17. 18. Batalha Abbey was built here in
commemoration of a Portuguese victory
over Spain. Construction started three
years after the battle, which occurred in
1385. The Royal Cloister is a Gothic
structure decorated in the Manueline style,
a uniquely Portuguese form which makes
use of complex decoration. Boytac, the
originator, completed the work in the 16th
century.

# 城堡及樓塔

Castles and Towers

羅馬時代的後期，騎士和基本領土的封建制度在歐洲確立了。擁有多數騎士的諸候們，相互競爭，而在領土較易受侵之地構築堅固的城堡的情形，大家也習以爲常了。

文藝復興之後，也開始修築華麗的城堡或宮殿，在中世紀修築的城堡，和教會不同的是，著眼於堅固的簡樸造型。這種城堡，在狹谷間築設高聳的城壁的同時，還必定加設幾座塔。加設塔樓的原因，除了守望的功能之外，也是權力的象徵，而樓塔是個易守不易攻的安全處所。在歐洲各村鎮的城門上，加設樓塔是非常普遍的，原因也是在於防守。

中世紀的義大利都市，雖有很多樓塔，但是這些樓塔多是都市名流爲相互競爭而私人修築的。這種修築的型態，與其說是防衛安全，毋寧說是誇耀財富及勢力；這種情形導致樓塔的高度越建越高，數目越來越多。附帶一提的是在佛羅倫斯（Florence）全盛期，全部有四百座的樓塔，據說當時的杜辛吉家（Toshngi）的塔甚至高達2500英呎（約80公尺）。

現在對小村鎮而言，各城鎮的樓塔已經失去了禦敵的實用性了，但是由遠處觀望時，醒目的焦點，仍是村落城鎮的象徵指標。

Late in the Romanesque period the knights, and the feudal system of land ownership which they upheld, came into being. Lords who were able to control many knights frequently fought each other for territory, and built castles to defend their home country.

After the Renaissance many beautiful castles and palaces were built, but before this time, in the middle ages, the castles were built for strength, and unilke the churches were of very simple design. These castles were invariably built on narrow, high ground, and fitted with any number of towers. These towers were not merely lookouts, and certainly not simply symbols of power, but rather were built to be difficult to attack, easy to defend, and in a position which offered the greatest possible safety. In many European towns towers stand on the wall above a town gate, and this was also to defend the entrance to a town.

In the medieval towns of Italy many towers are to be found, but these were usually built by the city nobility over their own homes, in the spirit of competition. Rather than for defense, these towers were built more as symbols of prestige, and as a result grew higher and more numerous as time passed. At its height, Florence had as many as 400 towers, with the Toshingi family tower reportedly rising 250 feet (80 meters).

In most towns and villages, the many types of castles and towers have long since ceased to function as defenses against invaders from outside. Instead, as structures easily seen from afar, they act as symbols, and important design elements, of the towns over which they stand guard.

1. Hautefort Castle, strategically situated on a hill in Perigord, France. The existing castle was completed in the 18th century, and is thus more like the royal palaces of the Loire Valley than a rampart.

2. A renaissance castle built overlooking a poor Andalusian town. Lacalahurra, Spain.

1.　位於山丘上，具戰略性的歐特富爾城（法國，柏甘蒂地區）。現在的城堡是在十八世紀時完成的，所以與其說是要塞防守，不如說是羅瓦爾河地區的皇家宮殿。

2.　建於文藝復興時代，俯視貧窮的安塔露西亞村落的城堡。拉卡拉弗拉（西班牙）

3.　卡納弗恩城堡（英國，威爾斯）。圍繞著大中庭的城壁外的樓塔，是各自獨立且具防禦作用的

4. 在米德爾布魯夫高85公尺的塔。（荷蘭）

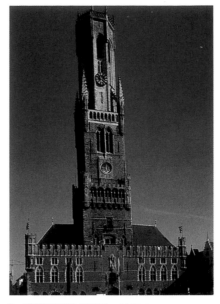

5. 紡織業者的辦公室及鐘樓。布爾日（比利時）

6. 面對坎波廣場而建的市政府建築，高88公尺的樓塔。錫耶那（義大利）

7. 建在村鎮口的城堡上的樓塔。羅騰堡。（德國）

8. 位於朗根布魯克的霍恩洛耶城內美麗的中庭，是原建於十五世紀，而在十七至十八世紀時重新擴建。（德國，霍恩洛耶）

9. 現在僅存的塔，是中斷了的50公尺高和100公尺高的斜塔，但在十二世紀末時，至少有194座塔林立於此。波洛尼亞。（義大利）

10. 建於民宅庭院中的樓塔。卡烈納古（法國，卡露西地區）

11. 聖上特利奴教堂獨立建造的木塔。翁弗勒（法國，諾曼第地區）

3. Caernarfon Castle in Wales, England. The walls of the outer rampart surround a large bailey (court yard), and stand as independent defensive positions.
4. The 85 meter tower in Middleburg, Holland.
5. Cloth Hall and the belfry. Brugge, Belgium.
6. The 88 meter Torre del Mangia at the Palazzo Pubblico, facing the Piazza del Campo. Siena, Italy.
7. The tower above the town gate. Rothenburg, Germany.
8. The beautiful central courtyard of Hohenlohes Castle, in Langenburg. The 15th century castle was expanded in the 17–18 century. Hohenlohe, Germany.
9. Presently, only this inclining tower of 100 meters and a broken one of 50 meters remain, but at the end of the 12th century there were as many as 194 towers of various height. Bologna, Italy.
10. A tower in the garden of a traditional house. Carennac, Quercy, France.
11. The free standing wooden steeple of Caterine Church, Honfleur, France.

# WINDOWS

● 窗戶

歐洲的房屋，街道的樣式，會因地區、國家等特性分明地表現出來。這個原因是除了建築材料和建築方法不同之外，實際上，最大的因素是令人意想不到的窗戶。在英國（England）（及愛爾蘭（Ireland））。地區，隨著所謂的喬治亞（Georgian）的古典建築型式的普及，十七世紀後半開始有的上下滑動式框窗幾乎取代了中世紀的橫向拉窗（或是由歐洲傳入的文藝復興式窗戶）；但是目前歐洲的內陸諸國，一般仍是使用絞鍊式的左右開口型窗戶。並且，除了伊比利亞（Iberia）半島外的歐洲諸國，通常玻璃窗的外側還會加設左右開啟式木板套窗（近年來在義大利（Italy）等地，設置捲場式百葉窗），但是在英國（England）及伊比利亞半島（Iberia）地區的窗戶，幾乎是不加設這種木板套窗的。而且，在西班牙（Spain）和葡萄牙（Portugal）地區，往往在一樓的窗戶都會加設鐵格欄，但是這種風氣似乎也成為國家的建築形態特色和特徵之一了。

其實在歐洲的村鎮及小聚落中，面對大馬路的房屋窗口，常裝飾得令人瞠目結舌。

在窗戶外面的植木鉢裡會種滿美麗的花朵，而窗口也會掛上綴著花邊的窗帘，除了這些，路上的行人也會不經意地看到窗櫺邊的桌上會擺有花瓶或其他裝飾品，也就是說路上的行人可以由窗口看見各家對室內裝潢所投注的心力。

其實窗戶或窗口的擺設之所以會如此講究，原因也是因為路人看見時的反應。因為路人經過時會不經意地往窗口看，而注意到窗戶的裝飾及擺設。在路人往窗內探望時，面對著馬路的窗口內，也同時有一雙寂靜地觀察著的眼神。若是屋主和路人正好是舊識，則可能還可以打打招呼，話話家常。窗口的裝飾及擺設，其實也是無言地暗示著「我住在裡面」，亦可以說是參與村鎮內交際溝通的一項證據。

When studying the distinctive architecture of the various regions of Europe, it is not only materials and construction methods which clearly differ. Surprisingly, the window has much to do with the unique appearance of a regional style. In England (and throughout Ireland) along with the appearance of classical Georgian architecture, which developed in the latter half of the 17th century, a new type of window, the vertically-lifted "sash," was introduced. It almost completely replaced the medieval window (or renaissance window, which had been imported from the continent) both of which swung open from the side. Even today, in Europe, most windows are hinged on both sides and pulled open. Moreover, throughout Europe, with the exception of the Iberian peninsula, windows have shutters on the outside, which are pushed open (although in Italy, in recent years, roll-up shutters are gaining popularity). Only in England and Iberia does one note an absence of shutters. In Spain and Portugal one sees metal lattices over windows, for security purposes, and this has become an architectural feature of these countries.

Windows which face onto streetfronts of European villages and small towns are often strikingly pretty. Outside are planters of flowers, while inside, the windows are not only draped with lace curtains, but, for the benefit of passers-by, often reveal window sills or a table upon which flowers and other ornaments have been placed. Interiors as seen through windows often show extraordinary care.

The style of a window is a response to the attention of the passerby. For whatever reason, people walking along a street tend to look unconsciously toward a window, and notice what is inside. At the same time, the resident has occasion to observe from his window what is happening on the street. If he happens to see an acquaintance walking upon the street, a greeting and conversation can take place. The well-cared-for window is a statement of its own, saying in effect : "I live here." In villages and small towns, the window has its own place in the life of the community.

1. 佩露吉（法國，柏甘蒂地區）

Perouges. Burgundy, France

2. 馬爾沃（葡萄牙）　　Marvão, Portugal

3. 馬爾沃（葡萄牙）　　Marvão, Portugal

4. 卡什特洛德維德（葡萄牙）　　Castelo de Vide, Portugal

5. 卡什特洛德維德（葡萄牙）．　　Castelo de Vide, Portugal

6. 加西翁（法國，普羅旺斯地區）　　Gassin. Provence, France

7. 馬爾沃（葡萄牙）　　Marvão, Portugal

8. 卡烈納古（法國，卡露西地區）　　Carennac. Quercy, France

9. 聖·西爾·拉波畢（法國，卡露西地區）　　St-Cirq-Lapopie. Quercy, France

10．伊奧（西班牙，加利西亞地區）Hío. Galicia, Spain

11．尤那威爾（法國，阿爾薩斯地區） Hunawihr. Alsace, France

12．聖・塞內里・洛・杰烈（法國，諾曼第地區） Saint-Cénéri-le-Gerai. Normandy, France

13．巴爾斯（西班牙，加泰羅尼亞地區）Pals. Catalonia, Spain

14．維羅納（義大利，威尼托） Verona. Venetia, Italy

15．巴爾斯（西班牙，加泰羅尼亞地區）Pals. Catalonia, Spain

16．威尼斯（義大利） Venice, Italy

17．克魯洛斯（英國，蘇格蘭） Culross. Scotland, Great Britain

18．卡瓦爾基（英國，蘇格蘭） Garvald, Scotland, Great Britain

19. 米滕瓦爾德（德國，巴伐利亞州）　Mittenwald. Bavaria, Germany

20. 米滕瓦爾德（德國，巴伐利亞州）　Mittenwald. Bavaria, Germany

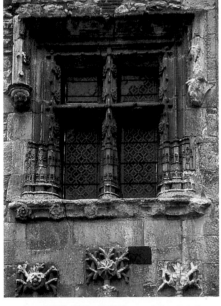

21. 卡奧爾（法國，洛特縣）　Cahor. Lot, France

22. 奧納斯（西班牙，安塔露西亞地區）　Osuna. Andalusia, Spain

23. 卡爾卡松（法國，洛特縣）　Carcassonne. Aude, France

24. 布勒隆‧安‧奧吉（法國，諾曼第地區）　Bouvron-en-Auge. Normandy, France

25. 吉爾布洛伊（法國）　Gerbroy, France

26. 吉爾布洛伊（法國）　Gerbroy, France

27. 塞布利亞（西班牙） Sevilla, Spain

28. 馬爾沃（葡萄牙） Marvão, Portugal

29. 威尼斯（義大利） Venice, Italy

30. 馬爾沃（葡萄牙） Marvão, Portugal

31. 迷你葡萄牙野外博物館。（科英布拉，葡萄牙）
Portugal in Miniature at Coimbra, Portugal

32. 聖提利亞那・德魯・馬爾（西班牙，坎塔布連地區）
Santillana del Mar. Cantabria, Spain

33. 聖・安德尼諾（法國，科西嘉島）
St.Antonino, Corsica, France

34. 奧德渥（法國，卡露西地區） Autoire. Quercy, France

35. 聖・雷奧（義大利）　San Leo, Italy

36. 拉文納（義大利）　Ravenna, Italy

37. 拉威洛（義大利）　Ravello, Italy

38. 萊切（義大利）　Lecce, Italy

39. 薩魯拉（法國）　Sarlat, France

40. 利克威洛（法國，阿爾薩斯）　Riquwihr. Alsace, France

41. 隆達（西班牙）　Ronda, Spain

42. 科爾尼亞（西班牙）　Coruña, Spain

# DOORS
● 門

當我們走在歐洲鄉村地方的村鎮時，往往可以在極為簡單的民宅聚落中，看見裝飾得非常漂亮的或是富麗堂皇的門。其實，家門或是建築物的入口，也可以說是通往對外道路的公共門面，相對於公共門面的設計，即使是小小木造洋房，屋門的設計也可以說是屋主藉此表達了自己的願望及尊嚴。

屋門的設計也有所謂的流行時髦樣式。在經濟許可的情形下，隨著時勢所趨地用更高級的材料等來替換原有的門及門邊飾物，也不是那麼不可思議的事。因此，中世紀的地方建築，常可看到古典式樣的屋門。但是，要以現代感的設計，更換掉舊門，是絕對不可能的。在現今的社會下，古老街道中被保存下來的屋舍，以及新建的仿古房屋，也是現代歐洲鄉間的時髦與流行。也因此現代有些國家，正試著那些不得不拆掉的古舊老屋的門等建築廢料，再度回收保管，並且在該地區重新修建、改建等時候，重新派上用場。

When walking through the villages and small towns of Europe one often sees very simple cottages and homes with stylish, even splendid doors. In fact, whether it be a small cottage or any other building, the door which opens onto a public street is looked upon as the face of the building, and its design a reflection of the owner's self-respect.

The design of doors is a matter of fashion. Where money permits, it is hardly strange to see a door and entryway which have been restored to a fine likeness of the original style. Thus one sees in medieval vernacular houses doors of classic design. What cannot be seen is an old door which has been replaced by a modern design. This is because preservation of traditional architecture has gained popularity, and because throughout European villages and small towns even new structures are made to look old. Moreover, when an old building must be destroyed, such elements as doors are removed and preserved, and used later when architecture in the region is remodeled, repaired, or built anew. Efforts to make use of such elements are taking place in many countries.

1. 于宰斯（法國，普羅旺斯地區）　Uzès, Provence, France

2. 露西昂（法國，普羅旺斯地區） Roussillon.
Provence. France

3. 露西昂（法國，普羅旺斯地區） Roussillon.
Provence. France

4. 露西昂（法國，普羅旺斯地區） Roussillon.
Provence. France

5. 班尼斯柯拉（西班牙，雷文特地區） Peniscola. Levant.
Spain

6. 佩露吉（法國，柏甘蒂） Perouges.
Burgundy. France

7. 巴爾傑姆。（法國，普羅旺斯） Bargème.
Provence. France

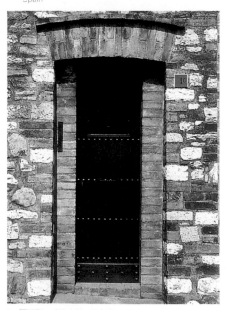

8. 阿西西。（義大利，翁布利亞地區）Assisi. Umbria, Italy

9. 加西翁（或，普羅旺斯地區） Gassin. Provence,
France

10. 安吉亞里（義大利，托斯卡納地區） Anghiari.
Tuscany, Italy

11. 阿西西。（義大利，翁布利地區） Assisi. Umbria, Italy

12. 聖提利亞那‧德魯‧馬爾（西班牙，坎塔布連地區） Santillana del mar. Cantabria, Spain

13. 古留意耶魯（瑞士，弗里堡州） Gruyères. Fribourg, Switzerland

14. 古畢奧（義大利，翁布利亞地區） Gubbio. Umbria, Italy

15. 露西昂（法國，普羅旺斯地區） Roussillon. Provence, France

16. 聖‧安德尼諾（法國，科西嘉島） St.Antonino. Corsica, France

17. 佩露吉（法國，柏甘蒂） Perouges. Burgundy, France

18. 杜布羅夫尼克（舊南斯拉夫，達爾馬提亞地區） Dubrovnik. Dalmatia, Former Yugoslavia

19. 費爾伯克（德國，霍恩洛亞地區） Vellberg. Hohenlohe, Germany

20. 聖‧塞內里‧洛‧杰烈（法國，諾曼第地區）
Saint-Cénéri-le-Gerai. Normandy, France

21. 奧爾塔‧德‧聖‧芬（西班牙，加泰羅尼亞地區）
Horta de San Juan. Catalonia, Spain

22. 加西翁（法國，普羅旺斯地區）　　Gassin.
Provence, France

23. 佩露吉（法國，柏甘蒂）　　Perouges. Burgundy, France

24. 桑‧斯薏露‧拉波畢（法國，卡露西地區）　　St-Cirq-Lapopie. Quercy, France

26. 聖提利亞那‧德魯‧馬爾（西班牙，坎塔布連地區）　　Santillana del Mar, Cantabria,
Spain

25. 聖提利亞那‧德魯‧馬爾（西班牙，坎塔布連地區）　　Santillana del Mar, Cantabria.
Spain

27. 克魯洛斯（英國，蘇格蘭）　　Culross.
Scotland, Great Britain

28. 奧爾塔・德・聖・芬（西班牙，加泰羅尼亞地區）
Horta de San Juan. Catalonia, Spain

29. 費雷（荷蘭）　　Veere, Netherlands

30. 卡特梅洛（英國，湖水區）　　Cartmel.
Lake District, Great Britain

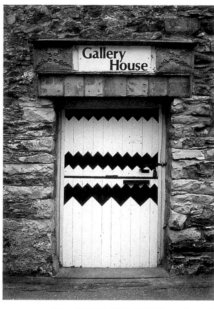

31. 卡特梅洛（英國，湖水區）　　Cartmel.
Lake District, Great Britain

32. 卡特梅洛（英國，湖水區）　　Cartmel.
Lake District, Great Britain

33. 布渥隆・安・奧久。（法國，諾曼第地區）
Bouvron-en-Auge. Normandy, France

34. 聖・佛羅倫斯（法國，科西嘉島）　　St. Florence. Corsica, France.

35. 布渥隆・安・奧久。（法國，諾曼第地區）　Bouvron-en-Auge. Normandy, France

36. 佩露吉（法國，柏甘蒂）　Perouges. Burgundy, France

37. 克魯洛斯（英國，蘇格蘭）　Culross. Scotland, Great Britain

38. 埃佛拉蒙特（葡萄牙，阿連特如地區）　Evoramonte. Alentejo, Portugal

39. 普魯紀（比利時）　Brugge, Belgium

40. 烏爾比諾（義大利，馬爾契斯　Urbino. Marches, Italy

41. 桑・斯懸露・拉波畢（法國，卡露西地區）　St-Cirq-Lapopie. Quercy, France

# PLANTS

●植物

對村落、城鎮而言，花花草草的存在可以沈澱撫慰人的心情，也可以增添街道的魅力。這些植物都是那些住戶們誠心誠意，親手照顧栽培的。在村鎮中的植物，常常大家所指的是那些門窗周圍可見的植物，這也是那些門窗背後居民用心的證明。

有時因怠於保養收拾那些花草時，雜草就會在馬路或建築物周圍蔓延生長，雖然這樣的景緻也可以讓人感受到鄉村樂趣，但決不能稱得上是具有情趣的美麗街景。

街道中的植栽和屋內的盆栽是一樣，也需要和周圍的環境相配合的。如果在街角的古井或是手押幫浦等周圍，以花草加以點綴，就可以表現出該地區居民對於自然環境的關懷及讚美。

植栽之中，有些長成的大樹，也可以發揮作用，成爲人們聚會的場所，或是可以達到遮陽蔽日的功能。此外，近年來完成的大馬路兩側，通常都會在道路的兩側種植行道樹。特別是夏日酷熱的內陸國家，在村鎮街市的道路兩旁，必定會種植行道樹。並且，若是那種筆直煞風景的道路兩側，種植大樹的話，就可以營造出安定感，而成爲一個令人安心的美好空間。

The plants and shrubbery of towns and villages calm the hearts of residents and add charm to the streetfronts. In reality, many of the plants and flowers seen in towns are lovingly maintained by local residents. The fact that many of the nicest displays of flora are seen around and at windows attests to this dedication.

Where a lack of interest has resulted in weeds growing up around buildings there is a worn appearance, and this in not at all to the benefit of a peaceful, pretty streetfront. Like flowers displayed in a home, the greenery of a town must be carefully arranged so as to complement the surroundings. Where the town's central well and pump have been carefully adorned with flowers, visitors will naturally sense the pride and love of the residents toward their town.

When it comes to large trees, they supply shade in the summer, and serve as a gathering place for dwellers. The large avenues of recent times are often lined on both sides with large trees, and many highways which leads to towns and villages of the continent are likewise sheltered from the hot sun by trees. Roads which are surrounded in this way by large trees fell secure, and can be a wonderful space through which to travel.

5. 佩拉斯特（舊南斯拉夫）　　Perast, former Yugoslavia

## 窗邊及建築物周圍的花朵
Flowers by windows and doors

2．阿爾科斯·第·弗倫特拉（西班牙，安塔露西亞）
Arcos de la Frontera, Andalusia, Spain

3．格根巴哈（德國）　Gegenbach, Germany.

4．布列斯洛（法國，奧佛涅地區）　Blesle. Auvergne,
France

5．尤那威爾（法國，阿爾薩斯地區）　Hunawihr, Alsace,
France

8．巴特·溫普芬（德國，霍恩洛亞地區）
Bad Wimpfen. Hohenlohe, Germany

9．尤那威爾（法國）　Hunawihr, France

6．安吉亞里（義大利，托斯卡維地區）　Angihiari, Toscany,
Italy

7．尤那威爾（法國）　Hunawihr, France

10．馬洛斯提卡（義大利，威尼托地區）　Marostica.
Veneto, Italy

12. 阿爾科斯·第·拉·弗倫特拉（西班牙， 安塔露西亞） Arcos de la Frontera, Spain

11. 佛羅倫斯（義大利） Florence, Italy

13. 尤那威爾（法國） Hunawihr, France

14. LY蒙契那斯（英國，威爾斯） Llanrhaeadr-ym-Mochnant, Wales, Great Britain

15. 安吉亞里（義大利） Angihiari, Italy

16. 費里安（英國） Veryan, Great Britain

17. 菲爾登茲（德國） Veldenz, Germany

76

# 裝飾街道的花朵
Flowers adorning the town streets

1. 尤那威爾（法國）　　Hunawihr, France

2. 尤那威爾（法國）　　Hunawihr, France

3. 佩露吉（法國）　　Perouges, France

4. 貝拉諾（義大利）　　Belluno, Italy

5. 聖‧馬洛（法國，布列塔尼地區）　　St-Malo, Brittany, France

6. 凱伊薩洛貝爾（法國，阿爾薩斯地方）　　Kaysersberg, Alsace, France

# 有大樹的廣場

A square with trees

1. 阿爾貝洛貝洛（義大利，阿普利亞地區）　Alberobello. Apulia, Italy

2. 西奧·德·烏魯蓋爾（西班牙）Seo de Urgell, Spain

3. 蒙特聖安傑洛（義大利，阿普利亞地區）　Monte Sant'Angelo, Apulia, Italy

4. 卡什特洛德維德（葡萄牙）　Castelo de Vide, Portugal

# 行道樹
A tree-lined drive

1. 托斯卡納地區（義大利）　Tuscany, Italy

2. 巴黎近郊（法國）　Near Paris, France

3. 托斯卡維地區（義大利）　Tuscany, Italy

4. 阿爾比近郊（法國）　Near Albi, France

5. 馬爾沃（葡萄牙）　Near Marvão, Portugal

6. 布拉托（舊南斯拉夫，科爾庫拉島）　Blato, Korčula, Former Yugoslavia

7. 基佛德（英國，蘇格蘭）　Gifford, Scotland, Great Britain

# STREET
# FURNITURE

●街飾

## 街燈

Street Lamps

　據說從古羅馬時代開始，都市的主要街道上，就已經設有街燈了。但是在西歐，即使是十八世紀的倫敦，夜間仍是盜賊出入的危險黑暗世界；到了開始設置煤氣燈，還是十九世紀以後的事。因此，所謂街燈的設計，無論看起來有多麼地古舊，其實都算是比較新的飾物了。歐洲各地鄉村及小城鎮，路燈完全設置完畢是最近的事，這個事實可以由古舊街道上，卻掛著計式樣新穎的街燈中得到印證。但是，和中世街景協調的不是現代感的設計，而是十九世紀以來的傳統式樣。另外，在歐洲的村鎮等鄉村小徑上的街燈，通常都是和以往的商店招牌一樣，是直接裝在建築物的牆壁上的。這樣一來，狹窄的街道便不會因為路燈而占用寶貴的空間了。

　　We know that the main streets of ancient Rome were illuminated by street lamps. But in western Europe, even in 18th century London, the nighttime streets were dark and dangerous, frequented by highwaymen and villains. It was not until the beginning of the 19th century that the first gas lamps were introduced. Therefore, no matter how antique these fixtures may appear, all are relatively new. In villages and small towns the appearance of street lamps is recent, and their modern design is often conspicuous against the old-fashioned architecture. But modern design does not go well with the old towns, and what one often sees on the narrow streets of villages and small towns are old-fashioned reproductions which, like the shopsigns, are often attached to surrounding buildings and therefore do not further crowd the narrow streets.

1. 阿西西。（義大利·翁布利亞州）　Assisi, Umbria, Italy

2．波爾多菲諾（義大利）　Portofino, Italy

3．安吉亞里（義大利，托斯卡納地區）　Anghiari. Tuscany, Italy

4．阿西西。（義大利）Assisi, Italy

5．波比（義大利，托斯卡納地區）Poppi. Tuscany, Italy

6．聖季米尼諾（義大利，托斯卡納地區）
San Gimignano. Tuscany, Italy

7．科托納（義大利，托斯卡納地區）　Cortona. Tuscany, Italy

8．佛羅倫斯（義大利）　Florence, Italy

9．奧魯維埃特（義大利，翁布里亞地區）Orvieto. Umbria, Italy

10. 維羅納（義大利，威尼托）　　Verona. Venetia, Italy

11. 安吉亞里（義大利）　Anghiari, Italy

12. 盧卡（義大利，托斯卡納地區）　Lucca. Tuscany, Italy

13. 古留意耶魯（瑞士）　Gruyères, Switzerland

14. 巴爾斯（西班牙）　Pals, Spain

15. 桑‧斯慧露‧拉波畢（法國）　St-Cirq-Lapopie, France

16. 聖‧塞內里‧洛‧杰烈（法國，諾曼第地區）
Saint-Ceneri-le-Gerai, Normandy, France

17. 科英布拉（葡萄牙）　Coimbra, Portugal

18. 烏比杜斯（葡萄牙）　Óbidos, Portugal

19. 巴特亞（西班牙，阿拉貢州）　Batea. Aragon, Spain

20. 隆達（西班牙）　Ronda, Spain

21. 波比（義大利，托斯卡納地區）　Poppi. Tuscany, Italy

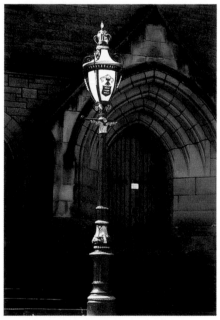

22. 皮布爾斯（英國，蘇格蘭）　Peebles. Scotland, Great Britain

23. 馬洛斯提卡（義大利，威尼托地區）　Marostica. Veneto, Italy

24. 安吉亞里（義大利，托斯卡納地區）　Angihiari. Tuscany. Italy

25. 貝拉諾（義大利，倫巴第地區）　Belluno. Lombardia, Italy

26. 威什薩（義大利，威尼托）　Vicenza. Veneto, Italy

# 水井及噴泉
Wells and Fountains

在許多中世的都市，飲用水的供給及確保是最重要的公共設施之一。首先，會將水井或湧泉處圍住以確保之，其次再於中心廣場處設置噴泉或飲泉處，接著才會設置周邊地區的水泉等。在夏季乾熱的地中海地區，水更是十分珍貴的。所以廣場的水井或是噴水泉，都是爲了帶給民眾便利歡愉所設計的。歐洲南方諸國，即使是小村落，也受到了古羅馬時代或者是伊斯蘭文化的薰陶，所以即使是在鄉間，也可以看到許多費心裝飾的噴泉或是飲水池。但反觀北方濕潤氣候的國家，水雖然也是生活上必須的物質，但是就不似南歐各國般視爲貴重品般珍視。所以在北方各地廣場中看到的，是供給飲用及洗濯用的水井或是幫浦，僅只是以實用爲目的設計。

In many medieval European villages and small towns the acquisition and supply of drinking water was a service of vital importance. First of all, wells and springs were contained and protected. Then, fountains and watering places were established in central squares. Later, fountains and watering places were built in other neighborhoods of the town. In the hot, dry regions of the Mediterranean, water was an extremely valuable resource in the summer months. Fountains and wells in central squares were, moreover, designed to give pleasure to the people. The villages and small towns of southern Europe, which developed under the influence of ancient Rome and the Islamic world, show a great variety of decoration in their fountains and watering places. In the north, while still important in daily life, the wells and watering places were not so invaluable as in the dry southern countries. In the north, the water supply in the central square was primarily for drinking and washing; most of the wells and pumps were therefore designed to be more practical than decorative.

2. 比恩察（義大利，托斯卡納地區）　Pienza. Tuscany, Italy

3. 佩露吉（法國，柏甘蒂）　Perouges. Burgundy, France

4. 維羅納（義大利，威尼托）　Verona. Venetia, Italy

5. 維羅納（義大利）　Verona, Italy

6. 尤那威爾（法國，阿爾薩斯地區） Hunawihr. Alsace, France

7. 查費洛休坦伊（德國，黑森林地區） Zavelstein. Black Forest, Germany

8. 隆達（西班牙） Ronda, Spain

9. 維羅納（義大利） Verona, Italy

10. 奧魯維埃特（義大利，翁布里亞地區） Orvieto. Umbria, Italy

11. 卡什特洛德維德（葡萄牙） Castelo de Vide, Portugal

12. 科西嘉島（法國） Corsica, France

13. 佛羅倫斯（義大利） Florence, Italy

14. 弗里堡州（瑞士） Fribourg, Switzerland   　15. 特里爾（德國，萊茵蘭－法爾茨州） Trier. Rheinland-Pfalz, Germany

16. 阿西西。（義大利，翁布利亞地區）Assisi. Umbria. Italy   　17. 勒普伊（法國，奧弗涅地區） Le Puy. Auvergne. France   　18. 里克威魯（法國，阿爾薩斯地區） Riquewihr. Alsace, France

19. 費爾伯克（德國） Vellberg, Germany

20. 卡米尼亞（葡萄牙）　　Caminha, Portugal

21. 古留意耶魯（瑞士）　　Gruyères, Switzerland

22. 維羅維（義大利）　　Verona, Italy

23. 貝加莫（義大利，倫巴第地區）　　Bergamo, Lombardia, Italy

24. 凱伊薩洛貝爾（法國，阿爾薩斯地方）　　Kaysersberg, Alsace, France

25. 勒普伊（法國）　　Le Puy, France

26. 埃佛拉（葡萄牙，阿連特如地區）　　Evora, Alentejo, Portugal

27. 巴杜莫（法國）　　Bargemon, France

28. 奧爾塔‧德‧聖‧芬（西班牙）　　Horta de San Juan, Spain

29. 埃施·斯洛·休魯（盧森堡）　Esch-sur-Sure, Luxembourg　　30. 杜姆（法國）　Domme, France

31. 安吉亞里（義大利，托斯卡納地區）　Anghiari, Tuscany, Italy　　32. 拉第科法尼（義大利，托斯卡納地區）Radicofani, Tuscany, Italy　　33. 馬洛斯提卡（義大利）　Marostica, Italy

34. 奧爾塔·德·聖·芬（西班牙）　Horta de San Juan, Spain　　35. 普勒洛斯（法國，奧弗涅地區）　Blesle, Auvergne, France　　36. 聖特堡（西班牙，加泰羅尼亞）　Santa Pau, Catalonia, Spain

37. 尤那威爾（法國，阿爾薩斯地區） Hunawihr. Alsace, France

38. 班拉第的傳統鄉村聚落。愛爾蘭。 Bunratty Folk Village, Ireland

39. 法弗沙姆（英國，肯特州） Faversham. Kent, Great Britain

40. 拉溫達布魯涅（瑞士） Lauterbrunnen, Switzerland

41. 聖德布利耶德（比利時） Zandvliet, Belgium

42. 拉溫達布魯涅（瑞士） Lauterbrunnen, Switzerland

43. 卡烈納古（法國，卡露西地區） Carennac. Quercy, France

44. 班拉第的傳統鄉材聚落。愛爾蘭。 Bunratty Folk Village, Ireland

# 時鐘

Clocks

　　設置在教堂或是城堡樓塔之上的時鐘，可能大家會認爲是一種附屬的裝飾品也說不定。但是，大家人手一只手錶之前，設在樓塔上的時鐘，是居民們得知正確時間的重要方法。

　　在發明鐘擺時鐘之前，一般都是使用日晷來計時的。雖然過去的人們的工作只限於白天，但是使用日晷鐘時是需要日照的，所以在日照較少的國家實在是沒有太大的助益。即使是現在，在地中海地區，日照充足的國家，仍可以看到日晷鐘的。

　　當十三世紀，機械構造的時鐘出現在歐洲修道院時，其實已經較中國晚了約二百年了。開始使用於修道院之原因，是因爲時鐘是修道院規律的生活上所必須的。

　　但是，爲了讓一般的民衆也能知道時間，所以將時鐘掛於教堂或是城塔之上，這就是鐘塔修建的緣故。但鐘塔修建也是在十七世紀後半，荷蘭物理學家惠更斯發明了鐘擺時鐘後的事。鐘塔修築的原因，除了是方便民衆知道時刻之外，也是因爲鐘擺時鐘的大鐘錘及掛鐘錘的長鍊子較有放置空間之故。

The clocks which are seen on castle towers and church steeples might appear to be mere accessories, but until the fairly recent advent of portable watches, the only way by which people walking on the streets of villages and towns could know the precise time was by these important fixtures.

Until the pendulum clock was invented, sundials were used. But even if we assume that all work was done during the day, sundials must have been of little use in the sun-deprived northern countries. However, in the sun-rich Mediterranean countries, one still sees a fair number of sundials.

The mechanical clock first appeared in Europe in monasteries during the 13th century, about 200 years after its invention in China. They were considered essential to the highly ordered life of the monastery. It was not until the second half of the 17th century, when the Dutch physicist Huygens invented the pendulum clock, that timepieces were fitted to church towers and castle gates, and common people came to know the exact time. It was not only for purposes of good visibility that clocks were built on high towers, though; this construction was well-suited to the heavy chains and weights necessary to run the pendulum and needles of an accurate clock.

1. 奧德渥（法國，卡露西地區）　　Autoire. Quercy, France

2. 曼圖亞（義大利，倫巴第地區）　　Mantova. Lombardia, Italy

3. 貝古奈亞（義大利，拉齊奧）　Bagnaia. Latium, Italy

4. 波比（義大利，托斯卡納地區）　Poppi. Tuscany, Italy

8. 羅滕堡（德國，巴伐利亞州）　Rothenburg. Bavaria, Germany.

5. 阿特拉尼（義大利，坎帕尼亞）　Atrani. Campania, Italy

6. 威什薩（義大利）　Vicenza, Italy

7. 貝拉諾（義大利）　Belluno. Italy

9. 維特爾博（義大利，奧齊拉）　Viterbo, Latium, Italy

91

10. 蒙特布魯布亞諾（義大利）　Montepulciano, Italy

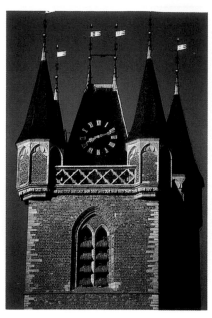

11. 斯露易斯（荷蘭）　Sluis, The Netherlands

12. 露西昂（法國）　Roussillon, France

13. 馬洛斯提卡（義大利）　Marostica, Italy

14. 于宰斯（法國）　Uzès, France

15. 威尼斯（義大利）　Venice, Italy

16. 聖・克洛瓦・安・嘉雷（法國）     St. Crois en Jarez, France

17. 佛羅倫斯（義大利）     Florence, Italy

18. 安吉亞里（義大利）     Anghiari, Italy

19. 科托納（義大利）     Cortona, Italy

20. 威什薩（義大利，威尼托）     Vicenza. Venezia, Italy

21. 特巴莫里（英國，蘇格蘭）     Tobermory. Scotland, Great Britain

22. 廷克爾斯畢爾（德國）     Dinkelsbühl, Germany

23. 克魯梅伊尤（義大利）     Courmayeur, Italy

24. 拉馬突耶爾（法國）     Ramatuelle, France

## 各式各樣的飾物

Adornments and Decorations

探訪歐洲的村莊聚落時，往往可以發現許多該國家或是該地區獨特的稀有飾品。

在英國的小鎮上，有些矗立在市場上的十字架（Market Cross），因地區及年代而產生各式各樣不同的形狀。但很可惜的是，現在這些都已成爲平台底座了。愛爾蘭彫著複雜浮彫的塞爾特十字架（Celtic Cross），雖然不太見於市鎮之中，但有些村鎮的馬路正中央，還是有樣式已難以辨認的大十字架豎立著。在法國則是用鐵打造成的十字架。在德國則常可以在路旁看見以彫花來裝飾的古舊葡萄榨汁機或手押幫浦。義大利的情形則是街角不時可以看到小的瑪麗亞像或浮彫。至於葡萄牙，則是常常可以看見瑪麗亞像被收藏在小祠堂般的祭壇上。西班牙的加利西亞地區（Galicia），是建立刻有聖經句子的石製十字架。這些傳說是以前塞爾特人（Celtic）在傳播基督教時所開始建立的。

這些可以增添歷史性的各種各樣的裝飾，對歐洲的各村落城鎮而言，是非常重要珍貴的文化遺產。像英國市集中的十字架，現在已經不具當時的功能了，但是它已經成爲村鎮保留的資產之一了；或者像是葡萄牙的瑪麗亞像，現在仍是村落中的日常信仰對象，而對一般民眾所重視；更有如德國的榨酒機械，現在已經不用使用了，但是過去的實用品，現在的功能卻僅成爲裝飾等，但是就是因爲有這些各式各樣的文化遺產，才使得街景更爲生動吸引人。

When visiting the villages and small towns of Europe one comes to recognize the distinctive decorations of the various countries and regions. In Britain, a good example is the market cross, which differs in design depending upon its age and location. Unfortunately, today all that is left of many market crosses is the original stand. The Celtic Cross of Ireland, with is complex relief work, is not often seen in Irish towns, although one might come across a huge, almost illegible specimen in the middle of a town street. In France, one sees holy crosses fashioned from wrought iron. In Germany, along roadsides, one comes across old flower-decorated wine presses and hand pumps. And in Italy, diminutive statues and reliefs of the Virgin Mary appear on many street corners. Statues of Maria, enshrined in an altar, are found in Portugal. And in Galicia, in Spain, stand stone crosses inscribed with passages from the Bible. It is said the first of these were erected by Celtic missionaries from Britain.

These relics, which add historical flavoring to streets and towns, are the invaluable cultural assets of European villages and small towns. Whether it be the market crosses of Britain, now no longer of practical use but nonetheless preserved by their towns, or the statues of Maria, still part of the religious faith of Portuguese villages, or the wine presses of Germany, once tools of daily life, now but well-maintained decorations, these various cultural artifacts are still adding to the life of towns in various ways.

1. 貝古奈亞（義大利，拉齊奧地區） Bagnaia. Latium, Italy

2. 佛羅倫斯（義大利，托斯卡納地區） Florence. Tuscany, Italy

3. 威尼斯（義大利） Venice, Italy

4. 威什薩（義大利，威尼托） Vicenza. Venetia, Italy

5. 肯特（比利時） Ghent, Belgium Liguria, Italy

6. 杜布羅夫尼克（舊南斯拉夫，達爾馬提亞地區）
Dubrovnik. Dalmatia, former Yugoslavia

7. 聖·卡西亞諾·第·帕尼（義大利，托斯卡納地區）
San Casciano dei Bagni. Tuscany, Italy

8. 廷克爾斯畢爾（德國，巴伐利亞地區）
Dinkelsbühl. Bavaria, Germany

9. 科托納（義大利，托斯卡納地區） Cortona.
Tuscany, Italy

10. 阿爾科斯·第·弗倫特拉（西班牙，安塔露西亞地區）
Arcos de la Frontera. Andalusia, Spain

11. 科托納（義大利） Cortona, Italy

12. 威尼斯（義大利） Venice, Italy

13. 聖·塞內里·洛·杰烈（法國，諾曼第地區）
Saint-Cénéri-le-Gerai. Normandy, France

14. 特里爾（德國，萊茵蘭－法爾茨州）　Trier. Rheinland-Pfalz, Germany

15. 佩露吉（法國，柏甘蒂地區）　Perouges. Burgundy, France

16. 威尼斯（義大利）.　Venice, Italy

17. 聖塔，菲奧拉（義大利，托斯卡納地區）Santa Fiora. Tuscany, Italy

18. 特爾博（義大利，奧齊拉）　Viterbo. Latium, Italy

19. 聖．卡西亞諾．第．帕尼（義大利，托斯卡納地區）　San Casciano dei Bagni. Tuscany, Italy

20. 普勒洛斯（法國，奧弗涅地區） Blesle.
Auvergne, France.

21. 比恩察（義大利，托斯卡納地區） Pienza.
Tuscany, Italy

22. 尤那威爾（法國，阿爾薩斯地區） Hunawihr.
Alsace, France

23. 聖提利亞那・德魯・馬爾（西班牙，坎塔布連地區）
Santillana del Mar. Cantabria, Spain

24. 聖提利亞那・德魯・馬爾（西班牙） Santillana
del Mar, Spain

25. 卡爾卡松（法國，洛特縣） Carcassonne.
Aude, France

26. 科托納（義大利） Cortona, Italy

27. 科爾庫拉島（舊南斯拉夫，達爾馬提亞地區）
Korčula. Dalmatia, Former Yugoslavia

1. 普魯紀（比利時）　Brugge, Belgium

3. 烏心杜斯（葡萄牙）　Óbidos, Portugal

4. 馬洛斯提卡（義大利，威尼托地區）Marostica. Venetia, Italy

5. 馬洛斯提卡（義大利，威尼托地區）Gubbio. Umbria, Italy

6. 安科拉（葡萄牙）　Ancora, Portugal

7. 雷・波・德・普羅旺斯（法國）
Les Baux de Provence, France

8. 奧納斯（西班牙，安塔露西亞地區）Osuna. Andalusia,
Spain

9. 特巴莫利（英國，蘇格蘭，馬爾島）
Tobermory. Mull Island in Scotland, Great Britain

10. 佩里格（法國，佩里格爾地區）　Perigueux. Perigord. France

11. 佩拉斯特（舊南斯拉夫）　　　　Perast. Montenegro, former Yugoslavia

12. 查費洛休坦伊（德國，黑森林地區）　Zavelstein. Black Forest,
Germany

13. 巴特・溢普芬（德國，霍恩洛亞地區）　Bad Wimpfen. Hohenlohe, Germany

1. 伊奧（西班牙，加西利亞地區）．Galicia, Spain

2. 古露格（法國，洛特縣）　Near Gluges. Lot, France

3. 布隆附近（法國，諾曼第地區）　Near Vron. Normandy, France

4. 雷·波·德·普羅旺斯（法國）　Les Baux de Provence, France

5. 勒普伊（法國，奧弗涅地區）　Le Puy. Auvergne, France

6. 雷·波·德·普羅旺斯（法國）Rignac. Lot, France

7. 里涅庫（法國，洛特縣）　Les Baux de Provence, France

8. 古留意耶魯（瑞士，弗里堡州）　Gruyères. Fribourg, Switzerland

9. 科隆馬庫諾伊斯（愛爾蘭，奧發利州）
Clonmacnois. Offeley, Ireland

10. 蓋魯茲（愛爾蘭，米斯州）　　Kells. Meath, Ireland

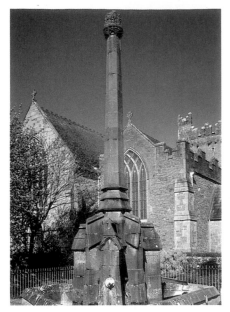

11. 阿達魯（愛爾蘭）　　Adare. Ireland

12. 波爾多菲諾（義大利，利古利亞）　Portofino.
Liguria, Italy.

13. 威茲列附近（法國，柏甘蒂地區）　Near
Vézelay. Burgundy, France

14. 梅瓦基塞伊（英國，康渥爾地區）
Mevagissey. Cornwall. Great Britain

15. 傑德巴拉（英國，蘇格蘭，波坦地區）　Jedburgh. Border. Scotland. Great Britain

16. 巴塞盧斯（葡萄牙）　Barcelos. Portugal

# 垃圾箱

Waste Baskets and Trash Cans

　　街道中必須設置垃圾箱，是在人們生活越來越豐富之後，垃圾箱也才隨之增加的。即使是現在，在一些貧窮國家，因為幾乎沒有垃圾，所以也根本不需要設置垃圾箱。即使是在歐洲，各村鎮等必須隨處設置垃圾箱，也是近代人們生活富裕了之後的事。因此，和古老街道協調的傳統垃圾箱式樣是根本不存在的。垃圾箱設計上的重要性，是在於街道間的協調，還有是為了不使拉圾箱立刻裝滿，而散落至周遭。行政當局在設計垃圾箱時，往往著重於顯眼，顏色鮮明，並且以低成本（含設計）為原則。看起來鮮明廉價的垃圾箱，無異於是對街道的一種暴力。比這個更令人難以忍受的是，垃圾箱溢出到周圍的垃圾。提到垃圾，先姑且不論過去的那些煙盒，餅乾等的包裝紙，現在還有一種開闢為野餐區的街道（特別是觀光勝地），所以小型垃圾箱，一下子就滿溢出來了。可以採行的應對之道，不是設置大型醜陋的垃圾箱，而是增加垃圾回收的繁率（必要時即使一天數次亦可）。

Waste baskets have become essential to towns as people have prospered and trash has increased. Even today, in the poorest countries there is practically no trash, and therefore little need of trash cans. The need for trash baskets in villages and small towns in Europe has come about recently, as a result of increasing prosperity. There is no such thing as a traditional trash basket in old towns, and thus no model for suitable receptacles.

　　The requirements for trash cans in villages and small towns are that they blend with the surroundings, and that they do not overflow, leaving the surrounding area covered with trash. Local governments seem to have decided upon a policy of making trash cans very showy, while investing as little money as possible in their manufacture (including design). Cheap, showy trash cans can even be said to do violence to the beauty of a town. The only thing worse is trash cans which are overflowing with garbage. In the past, the trash basket needed only suffice for containing cigarette cartons and candy wrappers, but today, especially in towns where tourists are apt to picnic, small trash cans can quickly become full. The only recourse is to collect accumulated garbage as often as necessary, even several times a day · but not to make trash cans larger.

1．理察摩得（英國）　　Richmond, Great Britain

2．康德・聖・馬爾（法國，盧瓦爾地區）　　Candes -St-Martin. Loire Valley, France

3．雷・波・德・普羅旺斯（法國）　　Le Baux de Provence, France.

4. 約克（英國） York, Great Britain

5. 貝加莫（義大利） Bergamo, Italy

6. 基爾肯尼（愛爾蘭） Kilkenny, Ireland

7. 烏比杜斯（葡萄牙） Obidos, Portugal

8. 皮布爾斯（英國，蘇格蘭） Peebles, Scotland, Great Britain

9. 卡尼約（安道爾） Canillo, Andorra

10. 卡拉聖伊提（西班牙，阿拉貢州） Calaceite, Aragon, Spain

11. 聖特堡（西班牙，加泰羅尼亞） Santa Pau, Catalonia, Spain

12. 卡什特洛德維德（葡萄牙，阿連特如地區） Castelo de Vide, Alentejo, Portugal

103

13. 班拉第的傳統鄉材聚落。愛爾蘭。
Bunratty Folk Village, Ireland

14. 馬洛斯提卡（義大利）　Marostica, Italy

15. 吉爾布洛伊（法國，皮卡爾地）　Gerbroy. Picardie, France

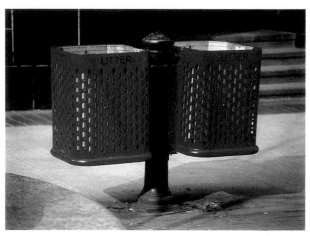

16. 古拉斯戈（英國，蘇格蘭）　Glasgow. Scotland, Great Britain

17. 聖提利亞那‧德魯‧馬爾（西班牙，坎塔布連地區）　Santillana del Mar. Cantabria, Spain

18. 夏蒙尼（法國）　Chamonix, France

# 電話亭
## Telephone Boxes

　電話亭，無論是哪裡，大致上都是同樣的規格。但是，最近以不漆油漆來節約維修費用，以及為防止長時間占用電話（或是避免電話機的破壞及其他，如將之當成廁所使用等），所以用鋁框鑲上玻璃，如此一來，大家就可以看清裡面，而防止上述的事情發生，也因此，這種缺乏感情的機能性產品，就進駐到各村鎮的古老市鎮中了。

　當然，電話機也是近代的發明產物，所以如果以中世的傳統半木造式或石造的設計來做為電話亭，也實在是有點愚蠢。但是，在古老街道中，像過去英國的紅色電話亭等，有細格窗狀的電話亭，似乎還較為相稱。

Telephone booths tend to be pretty much standard articles everywhere. But recently, cost cutting has led to less maintenance and poorer paint jobs. In order to discourage long telephone calls (and also vandalism, and even use of the box as a bathroom) more telephone booths are being built with glass and aluminum frames, which allow people to see in from outside. This is practical, but not aesthetic. More and more of these drab fixtures are appearing in villages and small towns.

　Of course, the telephone is a modern invention, so it is a little foolish to build telephone booths of medieval half-timber and stone designs. But it does seem that the old telephone boxes of Britain, with their small framed windows and red paint, are more appropriate to an old village or town streetfront.

1. 基爾肯尼附近（愛爾蘭）　Near Kilkenny, Ireland

2. 皮布爾斯（英國，蘇格蘭）　Peebles. Border Scotland

3. 埃佛拉（葡萄牙）　Evora, Portugal

4. 杜姆（法國，佩里格爾地區）Domme. Perigord, France

5. 曼圖亞（義大利，倫巴第地區）
Mantova. Lombardy, Italy

6. 道維爾（法國，諾曼第地區）
Deauville. Normandy, France

7. 卡奧爾（法國，洛特縣）　Cahors. Lot, France

# 長凳

Benches

為了讓疲累的徒步者，或者是讓大家有約會地
點，甚至是為了讓居民們有固定的集會地點，所
以衍生出長板凳的設置。可以有歇腳的地方固然
很好，但是也不是到處都可以設置這樣的長凳。
如果設在屋舍的窗口正前方，那麼就會不知不覺
地窺探窗口，然後再勿忙地錯開視線，所以是十
分不妥的。大致上，長椅都是設在廣場或是開放
的空間裡，如此一來，周圍特別是前方，就可以
有充裕的空間。也可以充分地欣賞到街景的美妙
之處。

因為如果有損毀時，就會十分破壞畫面，椅子
設計的重點，是在於耐久性；但是可惜的是，因
為太注重椅子耐久性的這個原因，反而無法顧及
休憩者的舒適性。

For tired pedestrians or those needing a place to meet
with others, or simply for locals who want to sit and talk
there are benches. It is certainly pleasant to have a bench
nearby when one needs a rest, but it isn't necessarily
good to have too many benches, either. Benches placed
right in front of people's homes, for instance, which force
one to look in upon others, are unwelcome. Most benches
are placed in squares, and other open areas, and all that
is necessary is that they have a far amount of open space
before them. If they offer a fine view of a lovely town,
then all the better.

The most important element in bench design is
durability. A broken or scarred bench is not a pretty sight
Unfortunately, and perhaps for that very reason, one does
not often come upon a bench which is also comfortable to
sit upon.

1. 科托納（義大利，托斯卡納地區） Cortona. Tuscany, Italy

2. 隆達（西班牙，安塔露西亞地區） Ronda. Andalusia, Spain

3. 費爾登茲（德國，摩澤爾地區） Veldenz. Mosel, Germany

4. 蒂羅爾地區的村落，奧地利。 Village in Tirol region, Austria

5. 塞普利亞（西班牙，安塔露西亞地區） Sevilla, Andalusia, Spain

6. 查費洛休坦伊（德國，黑森林地區） Zavelstein, Black Forest, Germany

7. 奧納斯（西班牙，安塔露西亞地區） Osuna. Andalusia, Spain

8. 貝拉諾（義大利，倫巴第地區） Belluno. Venezia, Italy

# 公共廁所

Public Toilets

　　現在，在歐洲的小村鎮聚落中，仍存有很多公共廁所。公共廁所是因有實際上的需要，應運而生的。並且，廁所要能確實清潔，預備上等的衛生紙，以達到乾淨舒服，這才是文明。可是並不是設置了公共廁所，就表示文明已經到達水準以上。最文明的，公認的一定是英國，但非常可惜的是還有很多鄉下地方，仍無法達到最文明的水準。在1870年，公共廁所最早開始由英國普及開來了，而且在之後不久，就發展出了現在的沖洗式馬桶。

　　公共廁所的設計，最好是以不太引人注目較好。在村鎮之中，最好是以方便使用為原則地設置在停車場之旁邊。但，若是有簡單易懂的指標，即使稍有不便也沒關係。

1. 馬洛斯提卡（義大利，威尼托地區）　Marostica. Venetia, Italy

Today, in many European small towns and villages there are public toilets. One does not appreciate the need for a public toilet until nature calls, and only then can one be truly grateful for its presence. A public toilet which meets ones need, i.e. is clean, stocked with good toilet paper, and in other respects satisfies its objectives well, is a mark of civilization itself. But the presence of public toilets is no guarantee of civilized facilities. The best public toilets are found in Britain, but unfortunately, there are many more countries which must still be deemed primitive. Britain was the first country to establish public toilets, and did so not long after the technology for the first flushing toilets was developed, in the 1870s.

　　The design of public toilets should be functional but not too conspicuous. From the standpoint of convenience, public toilets in villages and small towns should be built near visitor parking lots, where they can be used by tourists. If easy understood signs are provided, the toilets can be more out of the way.

2. 霍克斯黑德（英國，湖水地區）　Hawkshead. Lake District, Great Britain

3. 卡特梅洛（英國，湖水區）　Cartmel. Lake District, Great Britain

4. 勒普伊（法國，奧弗涅地區）　Le Puy. Auvergne, France

5. 普勒洛斯（法國，奧弗涅地區）　Blesle. Auvergne, France

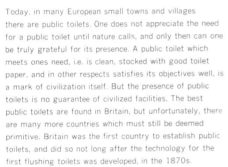

6. 盧夫勒（法國，卡露西地區）　Loubressac. Quercy, France

7. 地下的公共廁所。聖‧馬洛（法國，布列塔尼地區）　Underground toilet in Saint-Malo. Brittany, France

8. 地下的公共廁所。維羅納（義大利，威尼托）　Underground toilet in Verona. Venetia, Italy

# PAVINGS

● 舖裝路面及舖裝用的石材

自古以來在歐洲，就有用石頭來舖設路面的傳統。英文的street，德文的strasse，義大利文的strada等單字，都是源自於拉丁語（via）strata。古希臘的城市，雅典（ATHENS）的細長街道也並未加舖石，但若是看過龐貝古城遺跡，就可以明白古代羅馬都市，以舖石舖裝的馬路，還分成人行步道及車道。在古羅馬國滅亡之後，歐洲各城鎮開始舖裝道路，最早始於1185年的巴黎，接著是1235年，佛羅倫斯也開始了馬路的舖裝。到了1339年時，所有的馬路都舖裝上了舖石。此後到十六世紀之前，中世紀的歐洲各城鎮幾乎都已舖裝上舖石了，也因此，中世紀以舖裝道路為職的工匠們，也常常將藝術性融於工作中，展現出不凡的技巧。只是剛開始時，這種舖裝道路的目的是在於機能性而非藝術性。在街市的主要道路或廣場上加舖石材或進行舖裝，目的其實是為了防止下雨時的泥濘及天晴時的塵土飛揚，而且也可以讓往來頻繁的人車，走在固定的路徑上。另外，在往來頻繁的馬路上（特別是坡道），若未加設舖石，則貨車及馬車很可能會陷於進退兩難的困境中，實在有其不便之處，對於以市集為主的城鎮而言，也未嘗不是一種損失。

現在的歐洲，即使是在小村莊或聚落，也幾乎沒有一處不加舖石材的；但是無論是人或物的集散地，在鄉村地區的道路完全舖裝完成，也是最近的事。

在考量設計舖裝的要素時，最重要的就是和街景的搭配。在富有歷史性的村莊聚落，應該考慮使用自古以來就有的材料－石材。可能用大小的方型石或圓石等組合的這種表現方法，反而更能襯托出美麗的街景。但是若是舖石比周圍的民宅更為顯眼時，就有點流為本末倒置了。所以舖石的顏色，也是注意的要點之一。畫在石舖路面上禁止停車的黃線，往往也過粗並且太過鮮明。另一個引人注目的地方，就是舖石上醜陋的修補及損傷的痕跡。有些特地以手工修飾舖裝而成的美麗街景，但卻因埋設等工程，而將石舖破壞，之後並敷衍地了事地修補，這樣反而更凸顯醜陋的部份。

In Europe there is a long history of stone-paved streets. The English word "street," the German word "Strasse," and the Italian word "strada" are all derived from the Latin term (via) strata, which meant paved road. The maze-like streets of Athens, in ancient Greece, were unpaved, but in Ercolano and Pompei, the streets were laid with stones, and included separate paths for pedestrian and vehicle traffic. After the fall of Rome, the first European city to have paved streets was Paris, in 1185, followed by Florence, where paving began in 1235 and was completed in 1339. By the 16th century, most European cities had paved streets, and the artisans who laid the cobblestones had developed their work to a craft. However, the original purpose of cobblestones was functional, with the main objectives of paving streets and town squares being the protection of road surfaces from wheeled carts and carriages and the elimination of quagmires and dust. Where traffic was heavy—especially along steep road— wagons and horse carriages would block the way, causing impediments and incoveniences, and the town would lose its capacity as a marketplace.

Presently, there are virtually no towns in Europe, however small, that do not have paved streets. But with the exception of towns where traffic has always been heavy most of rural Europe has been paved quite recently.

When one considers paving as a design element it seems most important that the paving be in harmony with the streetfront. But when looking at the paving of historical town streets it is obvious that the stone, that ancient material, plays the leading part. It is the large and small square stones, and the combinations of round

1. 盧卡（義大利，托斯卡納地區）　Lucca. Tuscany, Italy

stones, which create the muted patterns which draw out the qualities of the streetfronts.

There is something amiss where the paved street is more conspicuous than the residences along it. Coloring is also a delicate matter. It easily happens that warnings painted on streets are too large, or loud in color. Damage and patching work on paved streets can also be very conspicuous and unsightly. Careless repair work on roads originally laid out with such care is really a waste.

2. 科爾多（法國，普羅旺斯）　Gordes.
Provence, France

3. 羅馬遺跡。埃努科拉諾（義大利）　The ruins of Rome. Ercolano, Italy

5. 古希臘遺跡。帕埃斯特杜姆（義大利）The ruins of Greek. Paestum, Italy

6. 佩露吉（法國）　　Perouges, France

4. 佩露吉（法國，柏甘蒂地區）　　Perouges.
Burgundy, France

7. 勒普伊（法國，奧弗涅地區）　Le Puy. Auvergne, France

8. 奧魯維埃特（義大利，翁布里亞地區）Orvieto. Umbria, Italy

9. 佩里格（法國，佩里格爾地區）　Perigueux. Perigord, France

10. 埃佛拉（葡萄牙，阿連特如地區）　Évora. Alentejo, Portugal

11. 奧米休（舊南斯拉夫，達爾馬提亞）　Omiś. Dalmatia, former Yugoslavia

12. 巴爾斯（西班牙，加泰羅尼亞地區）Pals. Catalonia, Spain

13. 烏比杜斯（葡萄牙）　Óbidos. Portugal

14. 克魯洛斯（英國，蘇格蘭）　Culross. Scotland

15. 亞德拉尼（義大利，阿馬爾菲海岸）
Atrani. Amalfi Coast, Italy

16. 格根巴哈（德國，黑森林地區）
Gegenbach, Black Forest, Germany

17. 聖提利亞那·德魯·馬爾（西班牙，坎塔布連地區）
Santillana del Mar. Cantabria, Spain

18. 塞普利亞（西班牙）　Sevilla, Spain　　19. 茲路特赫（德國，黑森林地區）　Siltach. Black Forest, Germany

20. 塞普利亞（西班牙）　Sevilla, Spain　　21. 奧爾塔・德・聖・芬（西班牙）
Horta de San Jnan. Catalonia, Spain

22. 塞普利亞（西班牙）　Sevilla, Spain　　23. 班尼斯柯拉（西班牙，雷文特地區）　Peniscola, Levant, Spain

24. 塞普利亞（西班牙）　Sevilla, Spain　　25. 塞普利亞（西班牙）　Sevilla, Spain

26. 馬洛斯提卡（義大利，威尼托地區）　Marostica. Venetia, Italy

27. 勒普伊（法國，奧弗涅地區）　Le Puy. Auvergne, France

28. 佩里格（法國，佩里格爾地區）　Perigueux. Perigord, France

29. 卡塞塔・維其亞（義大利，坎帕尼亞地區）　Caserta Vecchia. Campania, Italy

30. 桑・斯慧露・拉波畢（法國，卡露西地區）
St-Cirg-Lapopie. Quercy, France

31. 波爾多菲諾（義大利，利古利亞）　Portofino.
Liguria, Italy

32. 佩露吉（法國）　Perouges, France

# SIGNS
● 標誌

## 商店招牌
### Shop Signs

　　商店招牌，有和街道並排及和街道呈直角狀的兩種。其中最主要的就是，掛在建築物面牆壁上的看板，以及突出牆壁的立體招牌（牆側招牌）兩種。

　　以設計而言，特別需要注意的是第二種，突出牆壁的立體招牌（牆側招牌）。因為這種招牌過大時，會妨礙到視野，且有礙觀瞻。除此之外，若是在狹窄的巷道之內，就會讓人覺得前方全是招牌的錯覺。也因此，設計上的管理與控制就更為重要了。可以的話儘可能不要用現成的招牌，而用手製的招牌較能給人溫暖的感受，形狀大小也可以加以選擇，也能傳達出商店內的高尚氣氛，在街道中，這種商店招牌較能吸引人的眼光。

　　在鄉村城鎮之中，霓虹燈是大忌。如果無論如何都必須設計夜間的招牌時，可以在窗口或是櫥窗內也設個小小的招牌。

There are two kinds of shop signs, those which hang or parallel to the street and those which hang perpendicular to it. The main example of the former is the sign painted on the wall of a shop; the main example of the latter the protruding "sleeve" sign which hangs on the wall.

Of these the element which needs to be observed carefully is number two, the protruding sign which hangs at a perpendicular angle to the street. Does it protrude too far into the street? When looking down a narrow street one sometimes sees nothing but signs blocking the view. For this reason alone the regulation of town design is important. Preferable is the sign with homemade warmth, a sign in the old style, not something store-bought and certainly not anything too large; something that conveys well the character of the shop and at the same time complements the streetfront decor.

Neon signs should be prohibited in villages and small towns. When signs are required at night they need to be small, and set in either windows or glass displays.

1. 科爾多（法國，普羅旺斯）　　　Gordes. Provence, France

2. 頂克魯斯畢爾（德國，巴伐利亞州） Dinkelsbühl. Bavaria, Germany

3. 威茲列（法國，柏甘蒂地區）
Vézelay. Burgundy, France

4. 巴特·溫普芬（德國，霍恩洛亞地區）
Bad Wimpfen. Hohenlohe, Germany

5. 諾瓦耶·斯薏露·斯蘭（法國，柏甘蒂地區）
Noyers-sur-Serein. Burgundy, France

6. 利佛維埃（法國，阿爾薩斯地區） Ribeauvillé.
Alsace, France

7. 佩露吉（法國，柏甘蒂地區） Perouges.
Burgundy, France

8. 里克威魯（法國，阿爾薩斯地區） Riquewihr.
Alsace, France

9. 聖·馬洛（法國，布列塔尼地區） Saint-Malo.
Brittany, France

10. 埃斯波森德（葡萄牙，科爾塔，貝魯德）　　　Esposende. Costa Verde, Portugal

11. 古留意耶魯（瑞士，弗里堡州）　　　Gruyères. Fribourg, Switzerland

12. 策爾·阿姆·特威拉（奧地利，蒂羅爾地區）　　　Zell am Ziller. Tirol, Austria

13. 凱伊薩洛貝爾（法國，阿爾薩斯地方）　　　Kayserberg. Alsace, France

14. 米滕瓦爾德（德國，巴伐利亞州）　　　Mittenwald. Bavaria, Germany

15. 奧米休（舊南斯拉夫，達爾馬提亞）　　　Omis. Dalmatia, former Yogoslavia

16. 克爾索（英國，蘇格蘭，波坦地區）　Kelso. Border, Scotland, Great Britain

17. 布渥隆·安·奧久（法國，諾曼第地區）　Beuvron-en-Auge. Normandy, France

18. 穆斯第埃·聖特·馬利（法國，普羅旺斯地區）　Moustiers-Ste-Marie. Provence, France

19. 薩魯拉（法國，卡露西地區）　Sarlat. Quercy, France

20. 夏蒙尼（法國）　Chamonix, France

21. 哈靈頓（英國，蘇格蘭）　Haddington. Scotland, Great Britain

22. 古畢奧（義大利，翁布利亞地區）　Gubbio. Umbria, Italy

23. 維羅納（義大利，威尼托）　Verona. Venetia, Italy

24. 佩露吉　法國，柏甘蒂 Perouges, Burgundy, France

25. 勒普伊（法國，奧弗涅地區）　Le Puy. Auvergne, France

26. 穆斯第埃‧聖特‧馬利（法國，普羅旺斯地區）　Moustiers-Ste-Marie. Provence, France

# 村鎮的指標

Village and Town Signs

　　在西歐，趁著週末或是連續假期，大多數的人會到美麗淳樸的鄉間去渡假。所以為了這些前來度假的人，各地的村鎮都會在入口處掛上吸引人的指標。不過，太過於囉嗦的宣傳，是會得到反效果的。像是圖畫村或是花之鄉等簡捷的標語。但但是恐怕多數的觀光客，會先看好旅遊指南，並且先確認當地的地名，地點等各項要點，因此村鎮指標的句子及設計，似乎對觀光客前來的決定，不太有影響力。所以精心設計的指標，可能是對前來觀光者的一種誠心誠意的歡迎表現吧！更或者僅只是村民自我滿足的一項設計也說不定。

On weekends and during the long vacations there are many people who come out to visit charming villages and lovely small towns. At the gates to these villages and towns are very beautifully designed signs inscribed with town names and a greeting. Advertisements are considered tacky, and thus most signs offer simple catchphrases such as *Village Pittoresque* or *Ville Fleurie*. However, visitors to these towns are bound to have investigated their destination in a guidebook, and learned enough about it to care little what the sign looks like. The tastefulness of a village or town sign thus may not be very important as a design element. About the only people concerned with signs are the townspeople themselves, who want to show a particular face to visitors, or in some way satisfy a local wish.

1. 海伊高洛赫（德國，施瓦貝）　　Haigerloch. Swabia, Germany

2.

2.3. 貝魯內克（德國，黑森林地區）　　Berneck. Black Forest, Germany

4. 利佛維埃（法國，阿爾薩斯地區）　Ribeauville. Alsace, France

5. 桑‧斯慧露‧拉波軍（法國，卡露西地區）　St-Cirg-Lapopie. Quercy, France

6. 圖雷特‧斯洛‧魯（法國，普羅旺斯地區）　Tourrettes sur Loup. Provence, France

7. 布爾根（德國，摩澤爾地區）　Burgen. Mosel, Germany

8. 查費格休坦伊（德國，黑森林地區）　Zavelstein. Black Forest, Germany

9. 尤那威爾（法國，阿爾薩斯地區）　Hunawihr. Alsace, France

10. 波特（西班牙，加泰羅尼亞）　Bot. Catalonia, Spain

## 街道標誌
Street Signs

　在一些被漂亮地修復的了的古老街角或建築物的牆壁上，大概都會有一些特別設計的街道標誌，記著該街道的名字。記著街道名稱的標誌，是為了讓大家知道街道及地址的，可是在熟知小鎮各項事物的居民而言，這種街道標誌，實在是不太能發揮作用。因此，可能是最近大家開始關心街道設計及街景之後，才開始成為現在這種經過裝飾的街道標誌的吧。設計者以裝點街道的構思，竭盡技巧地來設計這種標示，但是和加設鋪石一樣，不要喧賓奪主地過度顯眼是設計的重點。

On corner buildings of old but renovated streetfronts one often sees specially designed street signs. The basic function of street signs is to indicate the address, so for town dwellers the design is not especially important. Rather, it seems likely that as interest has grown in the architecture of streetfronts, people have recently come to place greater emphasis on this element. And while it appears that much effort is being put into these decorations, it is to be hoped that they, like the paving stones of streets, will not be permitted to become too conspicuous.

1．穆斯第埃・聖特・馬利（法國）　　Moustiers-Ste-Marie. Provence, France

2．內爾哈（西班牙，安塔露西亞地區）Nerja. Andalusia, Spain

3．卡拉聖伊提（西班牙，阿拉貢州）Calaceite. Aragon, Spain

4．聖查格・德・康保斯特拉（西班牙，加利西亞地區）
Santiago de Compostela. Galicia, Spain

5．巴特亞（西班牙，阿拉貢州）　　Batea. Aragon, Spain

6. 維特爾博（義大利，奧齊拉）　Viterbo. Latium, Italy

7. 班尼斯柯拉（西班牙，雷文特地區）　Peniscola. Levant, Spain

8. 諾瓦耶·斯蕙露·斯蘭（法國，柏甘蒂地區）　Noyers-sur-Serein. Burgundy, France

9. 聖特堡（西班牙，加泰羅尼亞）　Santa-Pau. Catalonia, Spain

10. 穆斯第埃·聖特·馬利（法國，普羅旺斯地區）　Moustiers-Ste-Marie. Provence, France

11. 拉文那（義大利，艾米利亞－羅馬涅）　Ravenna. Emilia-Romagna, Italy

12. 錫耶那（義大利，托斯卡納地區）　Siena. Tuscany, Italy

13. 馬薩·馬利提馬（義大利，托斯卡納地區）　Massa Marittima. Tuscany, Italy

14. 巴爾斯（西班牙，加泰羅尼亞地區）　Pals. Catalonia, Spain

15. 加西翁（法國，普羅旺斯地區）　Gassin. Provence, France

# 地圖及其他標示

Maps and other signs

1. 唐克爾德（英國，蘇格蘭）　Dunkeld. Scotland,
Great Britain

2. 拉溫達布魯涅（瑞士）　Lauterbrunnen, Switzerland

3. 巴爾斯（西班牙，加泰羅尼亞
地區）　Pals. Catalonia, Spain

4. 威什薩（義大利，威尼托）
Italy　Vicenta. Venetia,

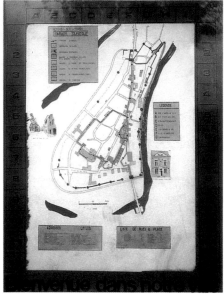

5. 諾瓦耶，斯慧露‧斯蘭（法國，柏甘蒂地區）
Noyers-sur-Serein. Burgundy, France

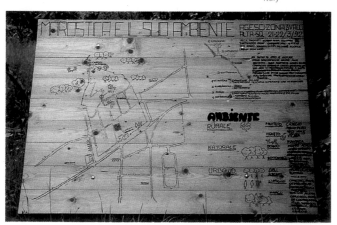

6. 馬洛斯提卡（義大利，威尼托地區）　Marostica. Venetia, Italy

7. 馬薩‧馬利提馬（義大利，托斯卡納地區）　Massa Marittima. Tuscany, Italy

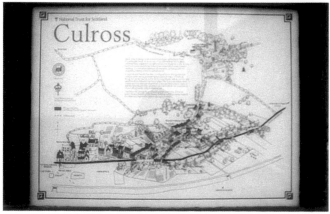

8. 克魯洛斯（英國，蘇格蘭）　Culross. Scotland, Great Britain

9. 凱伊薩洛貝爾（法國，阿爾薩斯地方）　Kayserberg. Alsace, France

11. 往豬博物館的指示標誌。巴特·溫普芬（德國，霍恩洛亞地區） Guide sign to Swine Museum. Bad Wimpfen. Hohenlohe, Germany.

12. 指示標誌。拉第科法尼（義大利，托斯卡納地區） Guide sign. Radicofani. Tuscany, Italy

13. 往市集廣場方向的指標。茲路特赫（德國，黑森林地區） Guide sign to market. Siltach. Black Forest, Germany.

14. 往露營地的指示標誌。勒普伊（法國，奧弗涅地區） Guide sign to campground. Le Puy. Auvergne, France.

10. 指示標誌。策爾·阿姆·特威拉（奧地利，蒂羅爾地區） Guide sign. Zell am Ziller. Tirol, Austria.

15. 「注意犬狗」（拉丁文）的標示。拉文洛（義大利） "Beware of Dog"(Latin) sign. Ravello, Italy

16. 汽車修理的招牌。威什薩（義大利） Auto repair shop sign. Vicenza, Italy

17. 公共電話的指示標誌。波比（義大利，托斯卡納地區） Telephone booth sign. Poppi, Tuscany, Italy

18. 市政府的招牌。馬洛斯提（義大利，威尼托地區） Town hall sign. Marostica. Venetia, Italy

19. 警察署的標誌。阿瓦隆（法國） A sign of a police station, Avalon, France

21. 學校的標誌。波比（義大利，托斯卡納地區） A sign of a school. Poppi, Tuscany, Italy

20. 市政府的招牌。達古蘭（法國，佩里格爾地區） Town hall sign. Daglan. Perigord, France

22. 美術學校的標誌。安吉亞里（義大利，托斯卡納地區） A sign of an art school. Anghiari, Tuscany, Italy

# SHOP
# FRONT

● 商店前方的擺設

選定在重要大馬路上或廣場開設的商店前方擺設，對村鎮中的街景而言，在設計上也是件重要的因素。對村鎮中大多數的居民而言，外出時的主要目的，就是購買（包括逛櫥窗），飲食等。對因街道美觀慕名而來的觀光客來說，購買當地土產，也是件十分有趣，而且樂於其中的事。所以商店前常是人群聚集，往來行人必經之路。具魅力的商店櫥窗，非但不會對構成街景的影響，相反地也會爲街道帶來無限的生氣及商機。而且引人駐足。

在商店門面擺設的設計重點，就是和街景相互調和。所以讓人感到街道的生氣勃勃是很重要的。將店名寫在商店的前方遮陽棚上或是掛上招牌，除了遮陽棚和油漆等，必須要注意到色彩的調和度，掌握街道和建築物的時代氣氛；也因此在這點的調和上是必要的。爲了讓人能感到生意盎然，不要偷懶是首要條件之一。不乾淨，或是剝落的商店門面，不但看起來不舒服，同時也破壞了街道整體的美感。

要維持商店門面的生趣，是面對大街及廣場的商店主人的責任。商店門面不能只是考慮到販售問題。如果街道整體感及氣氛遭到破壞，則相對地顧客也會跟著減少。所以在門面維持時，最重要的是每日開店及關門時的收拾整理。在店門口面積較大的門，若使用捲簾式鐵門，則店門一旦關起來之後，街道就顯得十分寂寞，彷彿之荒廢了一般。所以在有人通行處的店門擺設或玻璃門，最好不要關燈地維持白天的風貌。若是非關店門時，以普通平常的大門或大窗子，便可以了。

在歐洲的一些小聚落中，店家若是沒有懸掛招牌，則無異於一般民宅。而也有很多商店用的店門和窗戶，與一般家庭相同。也沒有陳列物品的空間；像這樣的小店，若能在房屋外側的牆邊展示商品，也可以增添街道的氣氛和生趣；應該是十分可愛討喜的商店門面的。

The facades of shops lining streets and squares are extremely important elements of village and small town architecture. For many dwellers and their neighbors, the main reason for going out on the town is to eat or shop (including window shop). For visitors to pretty towns, the purchase of souvenirs is likewise a major point of the trip. People invariably gather in front of shops, and the shop which is gaily adorned is never a detriment to the town. Instead, it attracts the public.

When considering the design of a shop front, the main question is whether the design harmonizes with the rest of the streetfront. Next, one will ask whether the shop design adds to the vibrancy of the street. The lettering of signs and other writing on the shop, as well as the positioning and coloring of signage, and how well these blend with the surroundings, are further matters of concern. Effort must be made to create a vibrant effect. Dust, flaking paint, or worse a shop that has a dirty appearance, are not merely ugly, but liable to ruin the appearance of the entire streetfront.

Owners of shops have a heavy responsibility to the residents of their town to maintain the quality of their shopfronts. If the appearance of a streetfront suffers, the number of visitors will dwindle and the business of the town will suffer. In this regard the handling of the shop front after hours and on holidays is of vital importance. When large display windows are shuttered it has a damaging effect on the ambience of the whole town. At least where pedestrians are common, it is important to leave glass displays open and lit. As far as possible, only normal sized doors and windows should be shut at closing time.

In many European villages and small towns houses cannot be distinguished from shops, unless signs are visible. There are many shops which have nothing more than regular doors and windows. There is likewise little space for the display of goods. In this case, goods may be displayed on the outer walls, with a positive effect on the image of the town. This is a lovable shop front.

1. 巴特‧溫普芬（德國）　　Bad Wimpfen, Germany

2．布魯日（比利時）　Brugge, Belgium

3．雷・波・德・普羅旺斯（法國）　Les Baux de Provence, France

4．佩露吉（法國）　Perouges, France

5．穆斯第埃・聖特，馬利（法國）　Moustier Ste Marie, France

6．拉文洛（義大利）　Ravello, Italy

7. 薩魯拉（法國）　Sarlat, France　　8. 莫斯塔爾（舊南斯拉夫‧赫爾茲格維那）　Mostar. Hercegovina, former Yugoslavia

9. 古留意耶魯（瑞士）　Gruyères, Switzerland　　10. 夏蒙尼（法國）　Chamonix, France

11. 杜姆（法國，多爾多涅地區）　Domme. Dordogne, France　　12. 聖提利亞那‧德魯‧馬爾（西班牙）　Santillana del Mar, Spain

13. 威尼斯（義大利） Venice, Italy

14. 克爾蒙（法國，阿爾薩斯地方） Colmar. Alsace, France

15. 威尼斯（義大利） Venice, Italy

16. 聖提利亞那·德魯·馬爾（西班牙） Santillana del Mar, Spain

17. 埃佛拉（葡萄牙） Évora, Portugal

18. 雷·波·德·普羅旺斯（法國） Le Baux de Provence, France

19. 拉文那（義大利） Ravenna, Italy

20. 科托納（義大利） Cortona, Italy

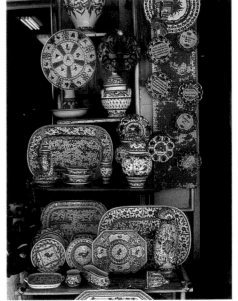

21. 阿爾科巴薩（葡萄牙） Alcobaça, Portugal

23. 維亞那・德・卡斯特洛（葡萄牙）　Viana do Castelo, Portugal

22. 布拉諾島（義大利，威尼斯）Burano. Venice, Italy

24. 雷・波・德・普羅旺斯（法國）　Les Baux de Provence, France

25. 古留意耶魯（瑞士）　Gruyères, Switzerland

26. 露西昂（法國）　Roussillon, France

27. 威什薩（義大利）　Vicenza, Italy

28. 諾瓦耶・斯慧露・斯蘭（法國，柏甘蒂地區）　Noyers-sur-Serein. Burgundy, France

29. 普勒洛斯（法國，奧弗涅地區） Blesle. Auvergne, France

30. 阿塞伊・德・利德（法國，盧瓦爾地區） Azey-le-Rideau. Loire, France

31. 翁弗勒（法國，諾曼第地區） Honfleur. Normarndy, France

32. 奧爾維埃特（義大利） Orvieto, Italy

33. 普勒洛斯（法國，奧弗涅地區） Blesle. Auvergne, France

34. 于宰斯（法國） Uzès, France

35. 法弗沙姆（英國） Faversham, Great Britain

36. 加西翁（法國） Gassin, France

37. 林渥薩哈（德國，黑森林地區）      Hausach. Black Forest, Germany

38. 杜姆（法國） Domme, France

39. 加爾富（德國，黑森林地區）      Calw. Black
Forest, Germany

40. 斯露易斯（荷蘭）     Sluis, The Netherlands

41. 露西昂（法國）     Roussillon, France

42. 拉馬突耶爾（法國）     Ramatuelle, France

43. 佩露吉（法國）     Perouges, France

44. 埃施·斯洛·休魯（盧森堡）　Esch-sur-Sure, Luxembourg

45. 基爾肯尼（英國）　Kilkenny, Great Britain

46. 普爾尼拉（西班牙）　Purunella, Spain

47. 普爾尼拉（西班牙）　Purunella, Spain

48. 翁弗勒（法國）　Honfleur, France

49. 雷·波·德·普羅旺斯（法國）　Les Baux de Provence, France

50. 布渥隆·安·奧久（法國，諾曼第地區）　Bouvron-en-Auge. Normandy, France

51. 巴頓·布拉德斯特（英國）　Burton Bradstock, Great Britain

# 露天咖啡及市集
Outdoor Cafes and Markets

在巴黎這樣的大都會裡，去喝咖啡已經不如從前般流行了，但是在一些徒步區內的戶外咖啡座（露天咖啡），現在仍是一樣受歡迎。在小村鎮中的咖啡座也是一樣受歡迎的，在氣候宜人的季節裡，歐洲人通常喜歡在戶外喝茶，吃飯。如果能將桌椅擺得適當，在人群聚集時，非但不會破壞街道的氣氛，反而可以賦予街道新的趣味。

特別能爲街道帶來朝氣的是，常常是固定日期的市集。平日感覺開闊的廣場，會擠滿各種露天商店，而且顧客們的喧嘩，就可以很容易醞釀出宛如祭典般的熱烈氣氛。這樣的露天市集，最常出現在氣候良好的南歐，北歐在夏天時，也偶而會有這樣的市集。

在廣場上佈滿，並排著商店的遮陽棚時，整個廣場的景觀也隨之改變。即使在廣場後有壯觀的建築物，但這時遮陽棚遮蔽了視線，同時大家的眼光也完全被五顏六色的商品所吸引了。露天市場固然是可以爲街道帶來新機，但有時卻也會對造成街道景觀破壞。因此在市集結束後，露天商店的棚子一定要收拾好。戶外咖啡座和露天市集，在石造的歐洲鄉間裡，可以增加街道不同於平時的熱鬧情景，所以適度地在郊外地區設置咖啡座或市集，可以在冷清的街道，多一點溫暖的人氣。

1. 廷克爾斯畢爾（德國）　　Dinkelsbühl, Germany

2. 波爾·格爾摩（法國）　　Port Grimmaud, France

In major cities like Paris it is no longer popular as it once was to sit in cafes. Nonetheless, many people still gather on cafe terraces, both in cities and small towns and villages. In good weather, Europeans enjoy dining and taking tea at an outdoor table. Well placed tables and chairs, and the pepole who gather at them, are never a detriment to the town setting. In fact, they add life and vibrancy to the avenues and squares.

Another feature which adds life to a town is the market, which is normally held on specific days of a week. The market square, which usually feels wide open, is packed with outdoor stalls at these times. There are many shoppers, and the feeling is not unlike the exhilaration of a festival. Outdoor markets are especially popular in southern Europe, where the weather is agreeable, but are also found in northern countries in the summertime.

With a multitude of pitched tents, the appearance of the square changes dramatically. Views of old buildings are blocked, and people's eyes are drawn to the colorful of the stalls. While markets lend life to the town, they also detract from the setting. Needless to say, when markets are finished the tents must be removed.

Markets and cafe terraces can lend the hard image of stone-built European towns an uncommonly showy atmosphere. Well planned cafes and markets can likewise lend a human warmth to towns which might otherwise seem cold.

3. 翁弗勒（法國）　　Honfleur, France

4. 翁弗勒（法國）　　Honfleur, France

5. 露西昂（法國）　　　Roussillon, France

6. 奧米休（舊南斯拉夫，達爾馬提亞）　　　Omis, Dalmatia, former Yugoslavia

7. 維羅納（義大利）　　　Verona, Italy

8. 里昂（法國）　　　Lyon, France

9. 威尼斯（義大利）　　　Venice, Italy

10. 貝沙魯（西班牙，加泰羅尼亞）　　　Besalu, Catalonia, Spain

11. 莫斯塔爾（舊南斯拉夫，赫爾茲格維那）　　Mostar, Hercegovina,
Former Yugoslavia

12. 曼圖亞（義大利）　　Mantova, Italy

13. 維羅納（義大利）　　Verona, Italy

14. 佩露吉（法國）　　Perouges, France

15. 貝古奈亞（義大利）　　Bagnaia, Italy

16. 巴塞盧斯（西班牙）　　Barcelos, Spain

17. 普魯紀（比利時）　　Brugge Belgium

18. 阿馬蘭特（葡萄牙）　　Amarannte, Portugal

19. 佩露吉（法國）　　Perouges, France

20. 莫斯塔爾（南斯拉夫）　　Mostar, Former Yugoslavia

21. 佩露吉（法國）　　Lucca, Italy

22. 普羅旺斯（法國）　　Perouges, France

23. 雷·波·德·普羅旺斯（法國）　　Le Baux de Provence, France

# UNSIGHTLY EXAMPLES
●錯誤的例子

1.

1. 停車、穿越馬路的汽車、林立的看板、電線桿札筆直的馬路，這些都是破壞街道景觀，使人不快的因素。要造成美麗具有魅力的街道，首先必須要排除路邊停車道，接著要整理突出建物壁面的招牌（特別是商品的招牌或是裝設著夜間照明燈光的看板）。電線也應該要埋設於地面之下。並且更進一步的，在道路的盡頭設置拱門等視覺上的擋牆的話，就可以輕易營造出令人安心的商店街氣氛。薩魯拉的村鎮盡頭。（法國）。

2. 照片右邊的電線桿架了很多電線，實在是非常地礙眼，雖然這電線桿已經下了工夫地沒有設在地面上，而是附著在建築物牆壁上，但仍是有礙觀瞻。像這樣有礙觀瞻的電線，對於深具歷史性的城鎮而言，無疑地是一大損害。（班尼斯柯拉，西班牙）。

3. 大型的招牌無論如何都是遮蔽視線的障礙物。但是，若是拆掉巨型招牌，就會露出這些荒廢了的建築物的醜陋壁面。在市府當局對這些建築物，進行修復或是重建計劃時，應該也看到這些招牌了吧！肯特（比利時）。

2.

1. Parked and passing cars. Clusters of signs. Telephone poles. Roads that cut through in straight lines...Such elements make the face of a town ugly and unpleasant. In order to bring out the full appeal of a town, through traffic must be eliminated, and signs which protrude from walls (especially product advertisements and lighted signs) controlled. Power lines should be buried underground, and, where possible, arches placed at the ends of streets, to create "eyestops", which promote a settled atmosphere within the town. Outside Sarlat, France.

2. The many power lines which cross the street from the telephone pole on the right side interfere with the view. The pole itself is actually affixed to the building. Nonetheless, the many power lines do much damage to this historical view. Peniscola, Spain.

3. At minimum, large billboards impede the view. But just removing signboards often leaves abandoned buildings and ugly walls exposed. Offices in charge of town planning have perhaps allowed billboards in lieu of repairs or rebuilding of such eyesores. Ghent, Belgium.

3.

4. 掛在廢棄了的房屋上大型的看板，造成了街道上荒疏了的感覺。再仔細一看，在這些廢棄屋外的牆壁上，還裝設著形狀優美的街燈，而且周圍房屋的窗台上，也用天竺葵盆景裝點著。如果能將大型看板拆除的話，應該也可以成爲一條迷人的街道。歐魯那茲，法國。

4. Large billboards on an abandoned building give a town an unkempt look. But if one looks closely, he sees a fashionable street lamp fixed on the building, and geraniums in planters by the doors and windows of surrounding houses. Unquestionably, removing these billboards would restore the charm to this small town scene. Ornans, France.

5.6. 大型卡車駛入小鄉鎮間，並停駐於其中，實在是將鄉村景觀破壞無遺。5的照片中，是運送啤酒到鄉間酒吧的卡車。霍克斯黑德（英國）。
在6的照片中的是停在具歷史性的羅馬式建築禮拜堂前的施工卡車。奧米休（舊南斯拉夫）

7. 街市中心區，即使整頓得十分美麗，但令人感到意外的是周邊（特別是田園和街市的交接處）的景觀卻是完全沒有整頓過。像這樣豎立在市鎮入口處的看板，就景觀上而言不只是難看，同時也破壞了市鎮給人的印象。薩魯拉（法國）。

5.6. Large truck traffic and parked cars can obliterate the views of villages and small towns. In 5., a large truck has arrived to deliver beer to a pub (Hawkshead, England.) In 6., a construction dump truck is parked in front of a historic Romanesque chapel. (Omiš, former Yugoslavia)

7. Even if the center of a small town is well kept, poorly planned surroundings (especially along the border between the town and its fields) are very common. Signboards like these at the entrance of a town not only spoil the view, but detract from the overall image of the town. Sarlat, France.

8. Where town preservation efforts have been directed upon old sections of a town the results are very pleasing to the eye. By establishing adequate parking outside of a shopping district, people who have fled for the suburbs will once again return to the center of town. However, the types of shop signs still tend to be too intrusive. If product advertisements can be limited, and old signs replaced with hand-crafted designs, not only will local shoppers return, but tourists will come from afar to enjoy the atmosphere. Manosque, France.

9. In Europe's beautiful towns and villages efforts have been made to remove unsightly telephone poles and power lines from main streets. But as this photo shows, back streets are often sacrificed in the attempt to save primary views. This is truly unfortunate. It would be well worth the cost to have poles and lines buried in these beautiful towns and villages. Carennac, France.

8. 街市中保存著古老風格的舊街道和建築物實在也是深具魅力的。如果能在這些舊街市的周圍增設停車場，會因購買空間的增加，而使流向郊區的購買人潮，再度回流到市區。但是，商店的各種招牌的稍嫌礙眼，所以，若是能把商品的看板和招牌重新整頓一下，再換成手工的別緻招牌，相信不只可以讓顧客回流，還可以吸引遠方來的觀光客。馬諾克斯（法國）。

9. 在歐洲有些美麗的村莊或城鎮，會費心地將不甚美觀的電線桿或電線，設置在馬路正面看不到的地方。但是一到馬路的背面去時，如這張照片中所示地，美麗的民宅後面常可以看到這些隱藏著的電線和電線桿。這樣的情形實在還是有點可惜，所以即使是要多花一點錢，還是將這些礙眼的電線和電線桿埋設至地下較好。卡烈納古（法國）。

# NEW DEVELOPMENTS IN VERNACULAR STYLE

## ● 傳統建築形式的新開發
—— 建築設計的新傾向 ——

現在歐洲多數的人們，都不喜歡現代主義的建築。可以看見一些戰後的建築，在修建當時排除了裝飾繁複的建築式樣，而改以機能簡單的現代主義式的建築，可是這些建築已經不再讓人感覺到新意了，反而給人一種無趣，與時代脫節的感覺。

人都喜歡新奇的事物。即使是關於建築，例如：英國在開始建築大型建築物的十八世紀以後，把各種式樣的建築及所有國家，所有歷史性的建築式樣完全融合，並巧妙地應用於不同時間，不同地方的折衷主義開始取代了古典復興式樣而盛行。

折衷主義時代同時也是各種機械開始登場的時代。反過來說，也可以機械的登上舞台帶動了經濟發展，使得大型建築的需求增加，所以折衷主義建築才能應運而生的。現代建築就是對重於裝飾的折衷主義建築的反動，「建築應該也要像機械般，不需要裝飾其機能性」，雖是將原因和結果互換的強硬想法，但是對新設計的構想上，仍是相當有力的說詞。

可是現今，不知道是大家對於現代主義建築已經厭倦了，還是如社會主義所言，那是歷史上的失敗，不過歐洲現在有很多人確實是認為那是一場失敗。不管如何，至少在村落和城鄉的街道上，確實是讓人感覺這是件失敗。為什麼呢？

那是因為現代主義的建築（例如：沒有屋頂的橫長大窗）和周圍的傳統建築無法調和，不只如此，現代主義的建築物相互間也難以達成協調。就像很多抽象彫刻作品齊集一堂的展覽會，反而給人一種混亂醜陋的感覺，即便是建築物本身都是藝術作品，但這樣的建築物組成的街市，反而容易流於混亂。而且很多新市鎮，是以現代主義的型式來修築的，當初是為求街道整體統一感的，但現在看來卻有過度單調之嫌，反顯醜陋。因此，若想以現代主義的設計來修築美麗舒適的街市的話，所有的建築物在設計上，必須有一共通的基本（可能需要以當地的傳統為基礎）設計，同時衍生出多樣的變化，但仍保持每一建築物本身原來的獨特性。

但是這樣的現代主義，其實是以傳統的本土形式而存在的。事實上，在歐洲有很多例子，就是以這種傳統本土形式開發形態，成功巧妙地營造出村鎮街市整體感的例子。

1. 現代主義建築的始祖，洛·科洛畢捷，建在馬賽的大型住宅區，為世界帶來潮流，對住宅計劃產生深刻的不良影響。這個巨型住宅區，即使是多加了許多細部的設計，但仍不太像是人類居住的房子。貝爾福（法國）。

For the most part, Europeans today are not fond of modern architecture. Buildings which were erected after the war were designed without architectural dressing or artistic flourish, as simple, functional structures. Today, these modern buildings inspire no interest or feeling, and appear drab and behind the times.

People like new things. In England, for instance, many large buildings were erected in the classical style until about the middle of the eighteenth century, when suddenly a variety of styles began to appear. Historical architectural styles from every age and country were used in appropriate situations, in the spirit of eclecticism. During this period, the so called Age of Eclecticism, a variety of machines were also invented. Put conversely, economic development created a need for a number of large buildings, which in turn gave birth to eclecticism with a number of architectural styles. Modernism rebelled against the decorative flourishes of eclecticism, however, insisting that architecture should be like machinery, namely functional. This powerful way of thought, i.e. that only means justify ends, made the modernistic approach an exceptionally forceful trend.

However, today it appears that society has tired of modernism, or perhaps that the modernism itself was the "historical failure" which is comparable to socialism. Many Europeans believe that the latter is probably the case.

What is undeniable is that this is reality when it comes to streetfronts in villages and small towns. Now why is this the case?

Modernistic buildings (e.g. roofless structures with horizontally long rectangular windows) do not blend well with traditional buildings, but that is not the only trouble. Neither do they blend well with other modern buildings. Just as an exhibit of random abstract art works soon becomes irksome and confused, so a streetfront comprised of modern buildings is easily create a confusing and repellent impression. Often when an effort is made to blend modern architecture, as in new developments, the result is a tedious repetitiveness, and again, the impression is ugly. If one really hopes to create a beautiful and pleasing street of modernistic architecture what is required is a common architectural element (most probably based upon the local materials and traditions) and at the same time some kind of variety (probably based upon design elements of artistic or artisan craft) in order to guarantee the individual buildings character, and the streetfront as a whole life.

But in this way modernism becomes the traditional vernacular style, and nothing else. In fact, across Europe the traditional vernacular style is being used to unify streetfronts in villages and small towns, and there have been many successes.

2. 集合整齊劃一的建築物而開發出來的新市鎮，即使每間房子都下了工夫，但是街道間卻缺乏吸引人的魅力。卡什特洛德維德（葡萄牙）。

3. 建築師奎藍‧泰勒在倫敦郊區所開發的新市鎮。是將十八世紀的新古典主義樣式（在義大利稱之爲喬治亞的古典建築樣式）的建築物，單純地將之結集在一起，雖然只是能感到簡單的群集，但是比起現代主義的建築手法，更能表現人們的情感，而且以商業的觀點而言，這也可以算是成功的建築。里士滿（英國，倫敦郊區）。

4. 建在木半造式房屋中的新建房屋。在歐洲，像毛山欅等堅硬的木材來源已經枯竭了，所以以往的半木造式建築的不可能再現了。加爾富（德國）。

5. 法國的國家事業之一，計劃於1963年，港口完成於1967年；自1968年開始，至今二十年的避暑渡假勝地－－馬林。雖然興建年代是在近代建築的全盛時期，但是現在來看，會覺得稍有官僚的庸俗氣息。照片是開始修築時，金字塔型的宿舍。（格蘭德‧莫特，法國）。

6. 建築家利卡爾德‧波菲爾所設計的修道院，在傳統，風土性的建築上，再以現代的修飾手法來完成的。安道爾。

7. 在西班牙，如照片所示地，將傳統的建築模式，擅加利用地放入新開發的建築中，看起來就十分自然。托克沙（西班牙）

1. The great proponent of modernism, Le Corbusier, built the Unité d'Habitation outside Marseilles, starting a movement which spread around the world and negatively influenced the development of housing architecture. Clever additions to the modern concept of large bureaucratie building design have failed to make it more suitable as a human dwelling. Belfort, France.

2. When built collectively, structures in new modern-style town developments fail to offer much attraction. Castelo de Vide, Portugal.

3. A new development on the outskirts of London, designed by the architect Quinlan Terry. This may seem like a mere borrowing of 18th century neo-classicism (called the "Georgian" style in England), but compared to modern architectural methods, it offers far more appeal to the eye, and is commercially successful, as well. Richmond, London suburb, England.

4. A new building amongst half-timbered structures. The disappearance of oak and other hardwoods from northern Europe has made impossible the construction of new half-timbered buildings. Calw, Germany.

5. Planning began in 1963, and the harbor was completed in 1967. From 1968, for the next twenty years, work continued on France's national plan to build a large marine resort city. The project was begun at the height of the era of modern architecture, and now appears somewhat uncouth, the result of overly bureaucratic planning. Grande Motte, France.

6. The monastery (N.D. de Meritxell), designed by architect Ricardo Bofil. Traditional and local design elements are arranged here in modern fashion. Andorra.

7. In Spain, examples of new developments which use traditional architectural styles—as in this photo—are fairly common. Toxa, Spain.

1. 位於庫拉林斯河和德威利德河河口處的梅利恩港，具有地中海村落的華麗氣氛。常有電影及電視劇，是以這個村落為背景拍攝的。

2. 村落中，禮品店精心設計的招牌。

3. 掛在村落入口處的導覽地圖。

4. 村落中心的綠色廣場上，建著義大利風格的廊柱建築（有並排柱子的建築物）。

## ●梅利恩港

這個宛如童話國度的村莊，座落於英國威爾斯（England, Wales）西北部，面對卡定甘灣（Cardigan Bay）的翠綠島上，規劃而成的人造村。這個村落的構想是來自威爾斯的建築家克羅．威利阿姆茲利斯利斯爵士（Sir Clough Willliams-Ellis），他年輕在義大利旅行時，發現了波爾特菲諾（Portifino）的魅力，這就是他構想的原由。他有一個夢想，就是在自己的祖國—英國，建一個足以與波爾特菲諾（Portifino）相媲美的村落，所以經過了幾年的搜尋土地，終於在1925年找到了這片美麗的土地。在這片土地上，他精心設計，採用了各式各樣的義大利的傳統建築風格，並且疆預定拆除的歷史建築，搬移到這片土上來，建造出精彩完美的村落。

威利阿姆茲利斯（Williams-Ellis）在純六十年以前的著作一〝英國和章魚〞（England-And Octopus）中，曾經寫著隨著汽車的普及，具歷史性的小村鎮和綠色的田園敲起了即將荒廢的警鐘，工藝家的威利阿姆茲．莫爾斯對二十世紀前的建築先驅，具歷史性建築及自然景觀，進行保存運動，這是一個推展保存運動的假想人物。之前克羅．威利阿姆茲利斯利斯爵士（Sir Clough Williams-Ellis）也曾經十分擔心梅利恩港（Port Merion）是否能取得拆除下來的建築材料，終於在1971年，這裡被指定為歷史性建築，並且作為歷史性建築的先驅，以此永久保存。而在利翌年，終於被加封騎士爵位，直至1978年，才結束他長達94年的人生旅程。

5. 威尼斯，尖頂式建築風格的拱門。

6. 有拱門的房舍。這拱門是通往村落的路。

7. 具地方色彩的木造小洋房的安排其實是很好的，但是貼著牆版的小洋房，與其說是地中海建築，不如說是義大利南路風格，更爲適切。

8. 巴洛克風格的尖形牆壁和壁上的壁畫，完全不是義大利風格，但是向外突出的窗戶，卻是維多利亞建築的特徵。

11. 這個村落中的小洋房，雖然塗著義大利式的中間色彩，但是這個頻色的組合卻不太像義大利風格。

9. 在小洋房的裡側，可以看見樓塔，在旁邊的樹也是義大利產的，因爲這個地點是在大西洋岸，屬於亞熱帶氣候之故

10. 這個圓頂建築，是建在一個由處就可以看見的顯眼的斜坡上。就是這個村落，是對自然環境加以利用而築成的。

12. 這個村落的建築，多是利用由拆除的舊建築材料來完成的。英國式小洋房的特徵在於房屋一定有巨幅的玻璃窗。

## Port Merion

This village in the northwest of Wales is like something out of a fairy tale, but it was constructed purposefully on this green peninsula on Cardigan Bay. The village was devised by the British architect Sir Clough Williams-Ellis, who in his youth traveled to Italy and was taken by the fishing village of Portofino. He spent the next few years searching for a place to build a village of equal charm in his own country, and finally, in 1925, came upon this beautiful piece of land.

Following this he set to work learning the necessary skills to build a traditional Italian village, and in the process of constructing it had buildings which were scheduled for demolition brought to the location.

More than sixty years ago Williams-Ellis, in his book "England and the Octopus," warned that the popularization of the automobile would lead to the abandonment of historical villages and rural lands, and with it led a revival of the movement begun by the artist William Morris two generations earlier to preserve historical buildings and the natural environment. Later, Port Merion itself was threatened with destruction, but finally, in 1971, these historical buildings, which exude so much humor and warmth, were set aside for permanent preservation. Williams-Ellis was knighted a year later. He lived to the full age of 94, passing away in 1978.

● Port Merion

1. Port Merion is situated between the mouths of the Glaslyn and Dwyryd rivers and has the showy atmosphere of a Mediterranean village. Many movies and TV dramas have been filmed here.
2. A decorative signboard above a village gift shop.
3. A guide map at the entrance to the village.
4. The green square in the center of the village sports Italian style colonnades.
5. Venetian gothic style arches.
6. A gatehouse with an arch. One passes under this arch to approach the town.
7. The placement of vernacular style cottages is superb. Only the weatherboard on the cottage here gives the place away as southern England, as opposed to the Mediterranean.
8. The baroque style gables and frescos are entirely atypical of England. However, the bay windows are very much a part of Victorian architecture.
9. In the back of the cottage one can see a campanile (detached bell tower). The tree beside it is typical of Italy, and grows here because this part of the Atlantic is in the temperate zone.
10. This domed structure, built above a slope, is conspicuous even from a distance. All the buildings of the town have been carefully arranged to make good use of the terrain.
11. The cottages of the town are painted pastel in order to create a southern ambiance. In fact, the coloring is not really typical of Italy.
12. Most of the structures in this town are built of materials taken from demolished buildings. This building shows the large glass windows typical of the English gothic style.

## ● 內爾哈

這是偶然在西班牙馬拉加(Malaga, Spain)附近的港都內爾哈(Nerja)郊外附近發現的,新開發型的地方色彩的住宅區。並排在小小的山丘上,安塔露西亞(Andalusian)特有的白色住宅,不規則地重疊修築而成,遠看彷彿是原來就有的村落。在西班牙,以地方傳統型式建築為設計要素的例子不只一處,最近的新開發村落常有此形態。特別是休閒地的開發,為了營造出風土性的氣氛,也經常將傳統建築形式加以現代的設計來修築。

1. 雖然每間房子都各不相同,但微傾的古老羅馬型屋頂和新近粉刷過的白色牆壁,卻是設計上的共通點。

2. 在門前小徑上,使用傳統聚落中的設計要素——拱門。

3. 由圍牆或門口的小徑上,可以看到門內房舍複雜的重疊。

4. 由各屋外陽台的重疊方式,可以知道這個山丘山的聚落是新開發的集合住宅。

## ● 科斯塔 · 斯梅拉爾德

在義大利的撒丁島(Sardinia, Italy)的東北部海岸線約30公里處,二十年前還是一片無人居住的荒涼地區,但現在卻被稱為科斯塔·斯梅拉爾德(Costa Esmeralda)是義大利目前最時髦的海濱度假休閒地。這裡的建築物是以撒丁島(Sardinia, Italy)或是地中海式的建築形式為主。在這建築物中,還同時有高級別墅、遊艇基地、休閒飯店、餐廳或是俱樂部,以及高級服飾店等等。其中在卡拉·第·渥爾貝地區的卡拉·第·渥爾貝店是超豪華的休閒飯店,雖是超級毫華的飯店,但是飯店整體卻會讓人感覺宛如古時候的村落,是因為飯店是以地方性建築的方式來修築的。

1. 看起來宛如是古代聚落的飯店外全景。因為沈浸在地方色彩的氣氛當中,所以在飯店的食宿費用也是十分驚人的。

2. 一個可以談話,又可以呼吸戶外新鮮空氣的陽台。牆壁外再塗抹上黃土。

3. 要讓整棟建築物看來更像是過去的房舍,牆壁上還特意留下一些深淺不同的斑點,但是仔細一看便不難發現,建築物本身其實是水泥構造物。

4. 右前方房舍壁上的塗抹方式等,看物以前的古老民房,但是停放在前方的休閒遊艇卻顯示出這是間飯店。

## ●庫里摩港

庫里摩港 (Port-Grimmaud) 是開發在法國南部聖特羅佩 (St. Tropez, France) 的一個人造的海上休閒區。這個小村落是由不規則形狀棧橋似的細長路地和小島,以運河及小橋相連結而成的。在這個村落之中,每一家的門口都可以進出遊艇。雖然全村的構想是有些類似威尼斯的感覺,但是這個村落的獨特之處是在於,這個村落徹底地表現傳統地方性建築特色。這個村落的設計是出自於法國阿爾薩斯的名建築家 A,斯波埃利 (A, Spoerry)。在設計當時的1960年代,正是近代建築最盛行的時候,但他卻執意地要建出地中海傳統建築,以至於我們現在才得以看到這個休閒區,也可以說是回歸傳統建築的先驅。

1. 以棧橋般細長的陸地連結漁村式的小別墅,在各房子的外側除了可以直接搭乘遊艇之外,也可以循著各連結陸地散步行走。聽說現在這些渡假小屋的身價都已達數億日幣了。

2. 除各間房子的牆壁和窗子的顏色之外,房屋的屋頂高度也各有些微的差距,因為這些所以看起來更像是自然形成的村落。

3. 由村落入口,通向中央廣場的橋上,所看見的運河旁的各房屋。雖然看起來樣式稍嫌混亂,但卻是不折不扣的傳統地方色彩。

4. 雖然村落內的廣場看起來彷彿也是古老村莊的一部份,但其實連廣場內的教堂都是新建的。

### Nerja

We came upon this is a new vernacular style housing development by chance outside the port town of Nerja, in the region of Malaga, Spain. Set atop a hill, this irregular-shaped gathering of traditional white Andalusian-style houses looks just like an old town when seen from afar. Not only here, but all over Spain, one can see where the vernacular style has been utilized in new construction projects. This is especially so in resorts, where the vernacular style is more and more being interpreted with modern concepts in ways which bring out the flavor and tradition of the surrounding locale.

### Costa Esmeralda

Until about twenty years ago a 30 kilometer stretch of the northeast coast of Italy's Sardinia island was uninhabited wilds. Now this area is known as Costa Esmeralda, and it is one of the most fashionable seaside resorts in Italy. Here all buildings are required to be designed in the vernacular style of Sardinia and the surrounding Mediterranean. Luxury villas and marinas, resorts and restaurants, night clubs and boutiques—all are in the vernacular. The Cala di Volpe Hotel, in the Cala di Volpe area, is a luxury hotel, but it could easily be mistaken for an old village so perfect in its vernacular finish.

### Port-Grimmaud

Port-Grimmaud, developed on the wetlands west of St. Tropez, in southern France, is a resort built on a man-made sea. The houses on these irregular-shapes spits and small islands are connected by maze-like canals and numerous bridges, and all can be reached by boat. While one can enjoy this town as a little Venice, it is more interesting still that the vernacular architecture has been so thoroughly recreated. The town was designed by A. Spoerry, the Alsatian architect, in 1960, at the height of the modern era. The architect was attracted to Mediterranean design, and the result was a new resort in the vernacular style—one of the earliest such examples.

---

### ●Nerja

1. The design elements of gently sloping roman tile roofs and freshly painted white walls are in perfect order, but looking closely one sees that the buildings themselves are different.
2. The traditional village design element of the arch is seen here above the street.
3. From a point on the road between the fence and the entrance one has a fine view of the town's houses.
4. From the layout of the houses with terraces one can determine that this hilltop village is a new development.

### ●Costa Esmeralda

1. A full view of the hotel, which appears exactly like a village from centuries past. In order to enjoy this authentic vernacular architecture one must pay an extremely high price.
2. A patio space for relaxation. One must be careful not to brush up against the walls, lest the genuine yellow mud paint soil his clothes.
3. In order to achieve an authentic appearance soils must be left in the walls. But if one looks carefully he will see that the underlying structure is in fact made of concrete.
4. In the painting of the walls and such the house

in the right foreground looks like an authentic traditional home. The leisure boat docked in front gives it away as a hotel.

### ●Port Grimmaud

1. The fishing village-style cottages are lined up along several strips of land which extend outward like narrow spits. All the cottages may be approached from the water by boat, or walked to over a network of bridges. Presently, the cottages are said to be worth several million dollars each.
2. Not only the colors of walls and the windows,

but the height of the roofs of each cottage is slightly different, giving the feeling that the village developed naturally.
3. Cottages around the canal as seen from the bridge between the entrance and the central square. Where the style is obviously mixed, the structures are unquestionably vernacular.
4. The square, which might be that of an authentic old village. The church in the background was also built recently.

# 創造舒適環境的處方箋

● 表一　歐洲各國的速度限制

| | 市區街道 | 普通道路 | 遠離市區之道路 | 快速道路 |
|---|---|---|---|---|
| 法國 | 60 | 90 | 110 | 130 |
| 奧地利 | 50 | 100 | 100 | 130 |
| 比利時 | 60 | 90 | 90 | 120 |
| 德國 | 50 | 100 | 130 | 原則上無限制 *1 |
| 愛爾蘭 | 48 | 88 | — | |
| 義大利 *2 | 50 | 110 | 110 | 140 |
| 荷蘭 | 50 | 80 | — | 100 |
| 西班牙 | 60 | 90 | — | 120 |
| 葡萄牙 | 60 | 90 | 100 | 120 |
| 瑞士 | 50 | 80 | — | 120 |
| 南斯拉夫 | 60 | 80 | — | 120 |
| 英國 | 48/64 | 80 | 112 | 112 |
| （日本） | 30/40 | 50/60 | 50/60 | 100 |

註：＊1.部份的高速公路有130km/h的限制。＊2.排氣量1300cc以上的車輛適用。

## 1.對於汽車社會應對

近年來歐洲各先進國家開始有了所謂「反都市化」的現象發生。因為以繁榮著稱的大都市（特別是工業都市）的人口已逐年減少，而在這之前人口大量流失的村落城鎮，近年卻開始湧進了大量的人口。

造成這種所謂「反都市化」現象的主要原因，是在於戰後自用轎車的急速普及。

因此在歐洲各先進國家對於這種自用轎車急速增加的處理對策，就是在小村落及城鎮中創造出快適，便利的環境。

### 速度的重要性要性

對交通而言，所謂的快適，便利性，最重要的就是速度吧！因為如果提高了速度，就可以節約時間，增加餘暇時間。同時也提高增加快適及便捷。如果速度增加成兩倍，則到同一目的的時間只需原時間的一半；花同樣的時間，則可以到達距離兩倍遠的地方。

因為面積是距離的平方，所以平均速度加倍的話，到達的距離也會加倍，則在同樣的時間內，可以橫越的面積，以地形和道路的延伸而言，可以成為平時的四倍。同樣地，如果速度增加成三倍時，就可以橫越過九倍的面積範圍。但是相反地，若是速度下降成一半時，同樣時間內僅能穿越原面積的四分之一。因此，即使平均速度只少了十分之三，但是能穿越的面積卻僅能達一半。

因此，對於目前這種汽車社會而言，創造一個快適便利的村鎮，是為提高車行速度（使車行速度不致降低），這樣的努力是十分重要的。如果能提高車行速度，就可以在較大的範圍內設置公共設施及增加就業機會等，因為若是能以這種快速穿越市鎮的方式來行進，則在村鎮間也不會比在大都市不方便。所以在創造村鎮間的便利的同時，確保連結各村鎮的路線及幹道是非常重要的。

在歐洲各先進國家，都已經作到了走出村鎮後，便能快速疾行的道路環境了。法國，德國，義大利，西班牙等歐陸國家，坐車的速度限制，在市區為50～60km/h，出了市區的普通道路為80～110km/h，遠離了市區的道路則是90～130km/h，在高速公路上的速度限制則是120～140km/h。在德國的快速道路幾乎沒有速度的限制，或是速度高達160～200km/h也是不足以訝異的平常事。為了增加車行速度，所以易於超車也是件重要的事。在法國，車道中央有超車專用三線道。在中央車道時，兩側都有可能有超車的情形發生，所以有時會覺得十分驚悚，但是不可否認地三線超車比兩線超車要容易得多了。在義大利的田園丘陵綿延的狹徑上，因坡度的關係上上下下的，所以若是漫不經心地追逐的話，是十分危險的。但是若是禁止超車，速度變慢，駕駛人又硬是超車，則反而會更增加危險。所以視區域及道路狀況來制定安全超車區，並仔細標示才是較安全的方法。

為了提高速度，所以另一件重要的事，就是儘可能地減少交通號誌。在村鎮外圍，即使是幹線道路，也沒有那麼大的交通量，視野若是還不錯，在普遍的十字路口就可以不必設交通號誌了。為了確保交通安全，保持十字路口的視野，以都市計劃來控制過多的帶狀開發（無秩序的沿道開發），遠比控制交通號誌更為重要。即使是交通流量大的地方，也不是非設交通號誌不可。在英國，大的十字路口都是間接迂迴的，所以也不需要交通號誌。即使是橫越通過的車子，也只需要減速或是停下來就可以了。因為不用等紅燈變綠燈的時間，所以在通過十字路口時，也可以較平時更為節省時間。並且，在英國，為了行人穿越而設置的交通號誌，在紅燈，汽車停下來後，立刻改為黃色點滅燈，就可以將等待號誌的時間減至最少。

不要有低劣的管理也是十分值得注意的。交通的取締及管理的目的，是為了預防事故的發生，但現在卻是以捉違規者為最大目的。應該是在超速就會容易發生事故的危險彎道上，取締危險駕車的；但是大家卻都在可以快速駕車，即使快速駕車也不致發生危險的大馬路上進行取締。這樣存心不良的惡性取締，反而會讓駕駛徒然地不安，造成安全駕駛的反效果。在這種惡性取締十分常見的西班牙，視不在危險的彎道上，不進行取締是件好事，以致於以汽車速度競賽般地如火如荼地展開。

### 控制帶狀開發

在發展成汽車社會的過程中，若是沒有作好計劃，則在幹線道路的材料場、工場、高速公路餐廳或是郊外型的購物中心等，就容易流於無秩序性地開發。對於要創造舒適的村鎮這件事而言，這種毫無控制的帶狀開發所帶來的害處，是怎麼強調都不為過的。

第一、每一個人都會注意到的缺點，就是醜陋。道路沿線的之廢棄場、材料廠、廉價廠房，或是那些掛滿招牌擁有巨大，透明圓形停車場的餐廳或是購物中心，這些不只是醜陋，更甚者是它們完全破壞村鎮的景觀。醜陋的帶狀開發區，不斷地由道路延線伸展至各村落之中，侵入了沿途的田園景緻，進而破壞了市區和田園的交界，完全地奪走了村鎮的景觀魅力。

第二個壞處是交通事故的增加，本來寬廣、視野良好與交通事故無緣的馬路，卻因帶狀發展，而導致十字路口等的能見度降低，在道路側邊加設各種設施，卻反而對人車的危險度增高。因為交通事故的頻繁，也使得道路的風貌也因而改變了。

第三個壞處，是為了因應交通事故而採取的對策，就是限制、降低時速或是加強取締，甚至是在路口設置交通號誌等，各項措施因而產生。為了改善交通事故而帶來的是行車時速的降低，以及如之前所提及的，間接損失了村鎮間的快適及便捷。

第四項壞處則是，帶狀發展使得村鎮的各重要商店客源流失，也奪走了村鎮的繁榮。這種帶狀發展除了導致交通事故的增加和車行速度降低，也使得村鎮中客源流失，又因發展區的沿路上開設了附有停車場的大型商店，連當地的消費者也都流向這些新興的商店。隨著這些新興商店急速的發展，更加速了村鎮本身商店的沒落。因為沒落而更導致村鎮看起來更顯得荒涼，更不用說魅力及舒適性。

第五項壞處，這樣的帶狀發展使徒步生活明顯困難起來，同時公共交通的效率也受到了影響。關於這一點，詳細地如下所記。

這種帶狀發展的害處，也可以說是阻礙村鎮發展，降低村鎮快適便捷，以及造成都市人口集中之元兇。在歐洲先進諸國，曾經也是帶狀發展過度的地區，但是在帶狀發展的壞處一再地被強調，加強認知之後，現在各地區都以嚴格的都市計劃來控制帶狀發展，結果現在大幅地提高各村鎮的便捷及舒適性。

雖然中世紀以來的美麗廣場，淪爲停車場，但是爲了維持中心商業區的繁榮，這也是無法避免的必要措施。法國，阿拉斯(Arras, France)。

對交通的便利及舒適性而言，最重要的就是速度感。英國湖水區的六米高速公路。

## 確保徒步生活

並不是有了汽車之後，人們便不需要步行了。村鎮的便捷快適，是使汽車在使用上更爲便利，還是人們的徒步生活也可以同樣方便呢？如果大家會有如此的猜測，其實也不爲過。換言之，如果能確實地達到，在自己的村鎮內徒步購物，到其他城鎮時則利用汽車等，兩種方式同時使用，並且爲了無法駕車的老者及孩童，確保公共交通，是大家最樂於見到的。這件事對於逐漸高齡化的先進諸國而言，可以說是當前最重要的課題。所以產生了兩項條件：1.將村鎮小型化及2.提高公共交通的率能。

但是阻礙上述兩項條件發展的最大因素，還是帶狀發展。那些建在帶狀發展區上，附有停車場的購物中心，原本就是蓋在村鎮中心區外的沿線道路上，而最初開始使用這些購物中心的，就是那些使用自用轎車的人，步行者要使用這些商店也是十分困難的。所以隨著帶狀發展區蓬勃發展，村鎮中的商店的沒落，更造成了徒步生活者的不便。但是問題不僅如此，帶狀發展區在村鎮的外環無計劃地過度開發，使得預定要設置的公共建設，如：公園、運動場、體育場、休閒娛樂設施等無法取得村鎮附近的外圍土地，結果導致民眾反而必需搭車才能前往遠離自己村鎮的休閒活動中心。現在歐洲各村鎮間、公園、運動場、休閒設施等增進村鎮舒適性的設施，都是設在村鎮的外圍土地上，民眾只需徒步，便可前往利用。控制帶狀發展，才有可能實現這種徒步生活的舒適便利環境。

連結各村鎮間的一般交通工具，就是公共汽車，但是阻礙公共汽車運行的原因，又是帶狀發展區。問題在於道路的行駛速度降低，沿途非必要性的停車等，使得到達目的地的途中花了過多時間。行駛時間的增加，同時造成需要量的減少，並且和相同次數的行駛相較之下，多花了許多不必要的燃料、車輛和駕駛費用，造成營運上的費用大幅增加。因需要量的降低，搭乘的收入也相對地減少；收入減少，但營運的費用增加，所以爲了不增加財政負擔，不得不提高搭乘的價格。可以提高搭乘的價格，相對地需要量就更形降低了，所以行車的時間增長，發車的頻率下降，價格的上揚（有時會因情況而廢止某些路線），就造成了這樣的惡性循環。對自己無法駕駛的人而言，若是不使用公共汽車，則唯一可以

使用的交通工具，就是計程車。在歐洲，即使是小小的村落，也都有公共汽車行駛往來；但是很多時候除了政府財政支援外，儘量控制帶狀發展區，保持村落的大小範圍，也是公共汽車能有效營運的重要原因。可是，有些地方還是除了計程車之外，沒有公共汽車可以使用。

## 停車場的重要性

對作爲交通方式的汽車而言，目的地也意味著順利停車的意思。要享受汽車社會的便利，就有兩點必要的條件：1.可以增加行車速度2.必需要能停車。所以在歐洲先進的國家裡，各村鎮通常也都具備了享受汽車便利的兩個條件。

其實在村鎮中，如果來訪的車子不多，在路旁就可以容易地停車了，不需要像在大都市中一樣辛苦地找停車位。但是，在村落城鎮中，前來觀光、散步、購物及吃飯等來訪的人一旦增加了，路旁停車的空間問題就浮現出來了。隨著村鎮中街景的保存，舒適性的提高，越來越具魅力的情形下，前來觀光、遊玩、購物等人口也不斷地湧入。所以換個方法，也可以說要創造村鎮的舒適性，就是要計劃停車場的問題。通常最開始遇到的問題，就是路旁停車的增加。對於小村鎮而言，路旁停車一旦增加，不但有礙觀瞻，還會造成交通的不便，另外在必要場合，反而無法停入車子，所以也會對當地居民造成使用汽車時的不便。但是沒有可以取代路邊停車的停車場時，也不可以就這樣禁止路邊停車。因爲如果禁止這樣的路邊停車，則這些小村鎮就沒有辦法發展下去了。即使有停車場，但在相當的距離之外時，還不如積極地同意短時間的路邊停車，還更能讓人覺得親切感。

對於路邊停車所採行的最差的對策之道，就是各商家在自家店舖的旁邊，建立自己專用的停車場。說這是最差的應對之道的第一個原因，在於醜陋。私人停車場的不美觀和將街道區隔成大塊狀的醜陋相互作用，更加深了其醜陋的印象。第二個原因是不方便。共同停車場和私人停車場，在使用上是大不相同的。購物或是吃飯，必要事先決定好在哪裡買，在哪裡吃等等，當然有很多時候沒有事前決定，而東走西瞧地看看之後才決定，也有別有一番樂趣在其中。如果使用專用停車場，就無法享受這種樂趣。另外，即使是事先

決定好了商家，也不定就只進一家店。買東西時，車子還必須再進第二家店，而且不符合經濟效益。例如：店裡平均會有三組客人時，若只預備三個停車位，就會將停車場塞得滿滿的。對於那些無法停到車子，而感到不便的客人，可能他們就會逐漸流失了。所以，假設要將塞滿的機率降到十分之一以下時，就必須預備五個停車位。若是這樣的商店有二十家，則就需要一百個停車位。可是，若是將這二十家商店的停車位，做成一個共同停車場時，只需要七十個停車位，就可以將擁塞的發生機率減低到十分之一以下。所以各商店的私人專用停車場不但不美觀，不方便，還十分沒有效率。

在村鎮中的公共停車場是免費，而且會儘可能靠近市中心區，一個停車場或是多不過兩、三個停車場，如果可以的話，都會儘量避開中央大道及廣場的必經之路，並且以設在大馬路及廣場上，無法直接看到的位置爲最理想。如果能控制村鎮間的帶狀發展區，則要設置這種理想的停車場也不是那麼困難。事實上，目前在歐洲各村鎮間，也幾乎都已依循這種方式設置停車場了。能夠如此設置停車的村鎮，才是能真正享受到汽車爲人們帶來的便利，同時也能創造，保持美麗的街道景觀。在各中央市區保存街景的同時，設置這種停車場，也可以說是環境計劃中重要的一環。

但是，若村鎮的規模稍大，則要建造出既能保持美麗街景，又同時要兼具便利性的停車場，實在不是件簡單的事。也有花大量費用，在市中心的地下建立大規模停車場的例子，但是經過多方籌措，大概也會多少犧牲掉一些街景或是便利性。其實，只要對於市觀沒有太過嚴重的妨礙，在路邊或是廣場有停車的空間的話，使用那些地點，也是常有的事。

但是有些車輛，無論如何停放都絕對會對街景造成影響的，那就是大型卡車。大型卡車和大型巴士不同，是在於大型卡車側邊下半部，架台下方是十分難看的，如果這樣的車子一停，是會完全毀了街道的整體美感的。所以如果在村鎮中必需設置大型卡車的停車場，則停車場就非得要完全隔離不可。

● 表二　國外觀光旅遊的支出金額（1986年）

| 國別 | 出超（單位一億US$） | 個人支出額（單位US$） |
| --- | --- | --- |
| 德國（前西德） | 206.7 | 339 |
| 美利堅合眾國 | 176.3 | 73 |
| 英國 | 89.2 | 157 |
| 日本 | 71.4 | 59 |
| 法國 | 63.8 | 115 |
| 荷蘭 | 44.3 | 304 |
| 加拿大 | 43.0 | 168 |
| 奧地利 | 42.2 | 558 |
| 瑞士 | 33.8 | 517 |
| 比利時 | 28.8 | 291 |

設立在索米爾（Saumur）村鎮中心區的露營區的指示標誌及露營區。

出處：OECD資料（J Christopher Holley, The Business of Tourism Pitman, 引用自1989起）
及'90～'91世界國政圖繪，國勢社。

# 2.和觀光共存

## 以休假爲取向的社會

在歐洲各先進國家，其實也可以說是一個「餘暇社會」。領薪水的上班族和勞動者的休假制度也十分落實而每個人的休假逐年增加。在現代的工作條件，雖然仍是週休兩天，但是已經有很多國家，在求職時就能保證年休三十天，連週休一併計算的話，可以年休約四十天。這麼長的假期要如何打發呢？這是歐洲人們最關心的大事。這也是因爲出生率及死亡率的降低，而人口進入高齡化的同時，所產生的「餘暇社會」現象。就西歐諸國來看，六十五歲以上的人口比例，在1910年時大約是6%，但是到了1960年，大約就已經達到了10%，到了1970以後，這個比例已升高到13%，因爲高齡者們具有較空閒的時間，所以也就對社會產生了大影響。特別是因二次石油危機，而造成失業率的突然升高，因經過失業經驗的洗禮，人們也更明白如何打發時間和享受休閒時間了。

現今，對於那些富裕先進國家的人而言，休假就是表示要離開家，到哪個遙遠的渡假勝地或觀光地去（通常都是國外）。也就是說「休假＝觀光旅遊（出國旅遊）」。或者是我們可以換句話說，歐洲的人們渴望旅遊的程度，就宛如在工作的一年間，他們所想的就只是休假。特別是一到了暑假，大家的話題就完全導向休假及去處等，例如：到哪裡去渡假了呢？如果還未休假時，問題就會改成：預定要去哪裡渡假呢？所以也常可以聽到一些不太好笑的笑話，因爲有些經營商店，較不寬裕（無餘裕出外渡假）的家庭，因爲顧及大家都去渡假，又怕自己無法去渡假會很丟人，於是就裝成要去渡假的樣子（實際上是沒有出門的），然後停止營業幾天，以掩人耳目。

如表二所見，就可以知道出國旅遊支出費用較高的國家，除了美國、日本，加拿大之外，其他完全都是歐洲的各先進國家。再看看個人支出金額，可以明白在歐洲各國中，特別是面積越小的國家個人支出額越高。在下面會更詳細地說明，但是在歐洲，即使說是出國旅遊，也可以僅用汽車就完成（只需燃料費就可解決）的情形很多，而且利用露營和帳篷等較爲便宜的設施投宿的情形，也十分普遍。

因此，在歐洲大家可以得到比付出金額更爲超值的休假，這就是休閒取向社會的實際狀況吧！

## 汽車的普及和露營帳蓬設施

曾經觀光旅遊（特別是長期的國外旅遊）是貴族或是富豪之家等特權階級，才能享有的奢侈生活。但是現在在各先進國家，對一般人家，也成爲司空見慣的渡假方式了。曾經也是特權階級才能使用的私人馬車，現在也普及了的自用轎車取代了。在1950年時，西歐各國的私家轎車合計起來不超過600萬台，但是現在全部已經遠超過一億台了。大約是每三個人就擁有一台轎車。例如：在英國國內的觀光客中，有百分之十九是使用私家轎車，百分之十二利用鐵路，使用固定路線公車的約有百分之七，利用觀光巴士的約有百分之三，使用出租汽車或飛機的各約百分之一，其他方式的共占百分之六，由此可知，以自用轎車作爲觀光旅遊交通工具的比例壓倒性地高出很多。但原來是島國的英國，海外旅遊時，大部份（百分之六十）是使用飛機的，可是也有百分之十三的人，利用大型輪船來載運而使用汽車。對於國境相連接的歐陸國家而言，即使是國外旅遊，大多數也都是使用私家轎車；例如在德國，和使用飛機的比例百分之三十六相較之下，使用汽車的比例仍超過百分之四十。

在休假曾是特權階級的特權時，休假時的住宿地點，通常是渡假勝地的別墅或是豪華飯店等，但是這對現在一般人而言是無法觸及的奢侈。特別是在休假時間較長，幾星期甚至是一個月以上時，不要說是豪華飯店，就算是觀光飯店、旅館或是民宿，這麼龐大的食宿費用，不僅限於在消費較低的國家，對一般家族旅遊者而言，這麼大筆的旅遊費用，也實在是難以負擔的。所以現在歐洲休假日數逐漸增多的今天，飯店或民宿的使用逐漸減少，而傾向利用膳食自理的設施。在這種設施之中，近年來益發流行的是大蓬車（像是附帶個小家庭的汽車，裡面有床，廚房，及衛浴設備等）或是露營車，以及有帳蓬設備的地方（有水電設備以及衛浴及淋浴等設施）。這種露營車或是大蓬車是長期租用或是一旦買下來，以後長期旅遊時還可以再使用，是可以很便宜地完

成旅遊的。通常在露營帳蓬設備的地方，也會有一些像是別墅般可以租用的投宿設備，稱爲固定式蓬車設備。例如：在英國國家公園，有一處稱爲「湖水區」的觀光點，其投宿的設施依統計來看，可投宿的床位約有十三萬七千九百個床位，其中包括旅館，汽車旅館及民宿等，全部包括在內也不過佔了全部比例的四分之一弱，僅約百分之二十四，其他的百分之七十六，都是可以自行炊事的設備。而且，大部份（約佔全體的百分之六十八）都是露營車及大蓬車（只是其中還有百分之三十是固定式的大蓬車）。

這種露營車或大蓬車的設備，分布在全歐洲，包括各有名的觀光勝地及大都市，但是這類的設備通常都是設在距離大都市有相當距離，不太方便的地方。而且在大都市之中，汽車的使用也不太方便，所以大部份的觀光客仍是會選擇投宿在飯店。不過，在西班牙等渡假勝地的露營及蓬車設備，通常也會包含泳池、高爾夫球場等休閒設備，所以以家族長期旅遊者爲訴求對象。但是這樣的蓬車及露營區的普及，除了那些較大的觀光勝地和有名的渡假地之外，含有這些設施的村鎮，的確是可以擴大促進村鎮的觀光事業。

## 汽車的普及和村鎮的觀光化

汽車的普及縮短了相隔遙遠兩地的距離，使得想觀光的人可以到較遠的地方進行訪問及探親。也就是說自家用車普及的結果，使得居住在較邊僻的人或是外來客人的往來，變得不再那麼稀奇。原來在觀光客較少時，來觀光也不過只是眺望景色，看看街景罷了；因此，即使是和地方上的居民有所接觸，也不致於會影響地方上的經濟生活（例如：農業）形態，對社會產生的影響也是微乎其微。但是隨著觀光客的增加，地方上居民的生活及社會也逐漸地改變了。

隨著觀光客的增加，觀光事業開始對地方經濟占有重要地位，而逐漸取代原來的農業經濟。特別是對於旅館，民宿等投宿設施的投資（特別是資金的借入），還有觀光道路等公共建設的投資等，使得觀光事業不再是業餘的副業，而爲了維持客源，也展開了激烈的宣傳及捲入了投資的競爭之中。社會上，家族及朋友間的往來，變成和觀光客間的互動，因和觀光客相處的時間變長，

戴波夏公爵城堡三百年紀念，到查茲渥斯宮殿來訪的人，在附近的小吃店排著隊。英國。

● 表三　國際觀光收入重要性之比較（1986年）

| 國別 | 國際觀光收支平衡<br>收入－支出（100萬US$） | 國際觀光收入<br>（100萬US$） | 國際觀光收入及<br>輸出之比率（%） |
|---|---|---|---|
| 西班牙 | 10,442 | 11,945 | 21.1 |
| 希臘 | 1,338 | 1,835 | 17.9 |
| 奧地利 | 3,197 | 6,928 | 17.7 |
| 葡萄牙 | 1,250 | 1,583 | 13.5 |
| 瑞士 | 862 | 4,240 | 8.9 |
| 義大利 | 7,095 | 9,853 | 8.6 |
| 土耳其 | 916 | 1,228 | 5.9 |
| 丹麥 | − 354 | 1,759 | 5.8 |
| 法國 | 3,197 | 9,580 | 5.2 |
| 愛爾蘭 | − 24 | 659 | 4.3 |
| 英國 | − 765 | 7,921 | 3.2 |
| 德國（西德） | − 12,838 | 7,826 | 2.6 |
| 比利時<br>（含盧森堡） | − 618 | 2,269 | 2.1 |
| 荷蘭 | − 2,524 | 1,906 | 1.8 |

出處：OEDS資料（引用自J Christopher Holley, The Business of Tourism, Pitman, 1989）。

進而對村鎮外面的世界也開始關心，並且增加見識。在家庭裡，太太的收入增加，也成為家庭生計之重要收入，女性的獨立性提高，家庭人口減少，教育水準也隨之提高。原來自給自足的農業生活，變成依附在都市周圍的商品及服務性都市生活，跟著前往都市謀職的人口也會減少，更甚者，因情況不同還有可能會有人口回流的現象。伴隨這種觀光化而來的社會及生活上的變化，會使得傳統的秩序變得脆弱，而造成伴隨經濟發展過程中，必然會產生的不良現象。在急速發展觀光化和休閒區域之後，像西班牙現在就產生了一些典型的，可能發生的問題。

## 觀光化的問題

　　歐洲的遊客，從以前開始就是追著太陽及溫暖的沙灘，而由北邊猛往南方跑。即使是冬季滑雪的時節，大部份的人也不去蘇格蘭或是斯堪地那維亞，而是到日照時間較長的阿爾卑斯地區等較南邊的地方。北方的人對於南部地方的憧憬，由德國文豪歌德的義大利紀行中，可窺得一二。另外很多在十九世紀末開始，有很多在工場或辦公室內工作的人，嚮往著健康的戶外休閒活動，並且蔚為風潮，同時曬成棕色的膚色也開始象徵著社會地位，所以接下來在法國南部，科特達祖爾的戛納、尼斯、蒙特卡羅、蒙頓（Menton）等各休閒渡假地區，就開始聚集了很多有錢有閒的人，這就是被稱為〝美好時代〞（belle epoque），而且建了很多典型折衷建築的豪華壯觀飯店的時代。

　　到了近幾年，曾經是權貴富豪們享受的「太陽，沙灘，海洋的休閒地」，也開始規劃開放給大量的一般顧客休憩，所以將價格開發得更為便宜，來吸引一般顧客，像西班牙的索爾海岸（Sol, Costa Del）就是逐漸往這個方向發展。加上因為經濟發展較遲，所以土地，建設費用，及工資等都比較便宜的西班牙，就因此而成為歐洲最大的觀光國，在海邊的現代化高層的飯店及休閒別墅等也漸漸在開發建設中。

　　如表三所示，為比較歐洲各國國際觀光收益的重要性，將輸出的國際觀光收入比率，依高低順序排列加以比較。由這個表格中，就可以發現西班牙的觀光收支平衡，收入額，及輸出比率，都

是佔第一位。另外，在表面上較高位置的國家，都是經濟發展的較遲的南歐諸國，或是奧地利，瑞士等擁有滑格渡假地的國家。義大利和法國，收支平衡而且還是黑字，但是那是因為這兩個國家不但臨地中海，還和阿爾卑斯山相連。而再看看英國，德國（前西德），荷比盧三國等北方工業國家，也有相當的國際收入，但是收支卻是完全的赤字，可以明白在觀光上，是出超的情形。

　　在1986年，觀光輸出佔21.1%的西班牙，已經比過去全盛期低落了，在過去西班牙佔有更重要地位，1970年時，西班牙的國際觀光收入比率曾經高達33%。西班牙最近觀光發展開始入低落的原因，不光是因為隨著經濟發展而輸出增加，更重要的是因為急速發展的觀光事業，污染了海岸、都市化的過程過於迅速，而且這種太陽，沙灘和海的休閒型開發，開始出現了新的競爭對手。由最近的時代週刊（1991年7月22日）中，就記著因為這種傾向，致使不只是西班牙，連法國的科特達祖爾也因而受到影響，導致一年間觀光收入銳減了百分之三十。

　　對於這種太陽，沙灘和海型態的休閒開發區而言，新的競爭對手，是什麼呢？最近典型的休閒型態開發，開始被新的開發型態所取代，這種新的型態，就是根據一些仔細周到的觀光市場問卷調查，而修建的大規模複合式的綜合休閒設施。除了周圍有可以栓快艇帆船的遊艇基地的飯店、休閒旅館、別墅等住宿設施之外，還有各式各樣的餐廳、咖啡館及時裝店，同時也有高爾夫球場、網球場、游泳池等娛樂設備的複合性的綜合休閒設施，常藉國際性的觀光資本家之手，建立在西班牙、土耳其、突尼斯等沿地中海岸地區。筆者也曾去過有這類休閒中心的突尼斯，對那裡留下了深刻的印象，因為不僅有完備先進的休閒設施，同時前往的旅客多是來自各先進國家。

　　在這麼完整的綜合設備之中，遊客無需離開，就可以享有所有的設備及做所有想做的事。只要具備了陽光，沙灘及海等條件，無論在哪一個國家，對休憩的內容都是不會有影響的。當然這些休閒地也是具多樣選擇性的，有一般大眾型的，也有少數財主渡假用的。只是這些問題的決定都只是在「價格」的變動上而已。換言之，哪邊都有，哪裡都可以建造具有的太陽、沙灘及海水的觀光賣點，而這其中的勝負關鍵就只是在於價格方面，如果交通費用在旅費中所佔的比率不太高的話，那麼隨著假期越長，物價及租金較為便宜

的國家，就較具競爭力。

　　其實，觀光業是低工資的，而且在七至八月的旅遊旺季時，對於歐洲先進國家的受雇者而言，沒有必要非工作不可的。因此，即使是在西班牙，觀光業者雇用的不是本地人，而是東南亞等地外出工作的勞工。對於投資者來說，觀光業或許是個富有利益的事業，但是對觀光業的工作者而言，卻是一項低收入的非安定性的工作。

　　加上因為觀光開發進而觀光客大量增加，接著所產生的交通等問題，造成了周圍環境的破壞，而引發當地居民生活上的問題。隨著觀光旅行的普及和大眾化之後，觀光客的服裝，常常會有上空或光是T恤，短褲，帆布鞋，就出國旅遊了。除此之外，觀光客的禮貌也越來越低落。因為是安靜，乏人問津的村落，所以若是有外國觀光客來訪，大家都會十分熱情地招待歡迎，但是現在這樣親切的畫面已不復存在了。當然，因為在好不容易的休假日，卻有各式各樣的外國人在自己家裡的附近搖頭晃腦，反而讓當地的人無法好好安靜地休息，所以這也可以說是觀光發展所帶來的負面義意！

## 村鎮的觀光化

　　觀光旅行是以「地點」和「事件」作為主要目的，再加上投宿地點的「設備」及「飲食」為次要目的，所共同組成的樂趣。對於採用露營和大蓬車式的旅遊者而言，「設備」和「飲食」是沒太大關係的。「場所」則又分成兩方面，一方面是海邊及山林，太陽，新鮮空氣等「大自然」，另一方面則是指可以表現出歷史，文化，藝術等的建築物，以及街景與有主題的公園等人為建築物。另外的「事件」，則是指傳統的廟會或是現代藝術的節慶，或是運動等活動。

　　如果說以往的旅遊是「太陽，海及砂灘」為目的的話，那麼現在「歷史遺產」則是旅遊的新目的。在歐洲有些有名的音樂祭，會以有名的街道為背景地舉行。換句話，我們也可以說歐洲可以用本身的文化歷史資產之旅，來取代並對抗傳統的陽光沙灘海洋之旅（就是以獨特性取勝，而不以價格來競爭），使之成為觀光的新主流。因此，近年來的新潮流之中，越來越受歡迎的觀光主題，也是本書的目的，就是殘存保留過去街景

每年五月初都會舉行的中世紀露天劇，稱爲Calendimaggio。

●表四

| | 人口 | 店舖的數量 | | |
| | | 小酒店及小茶館 | 食品行 | 雜貨及家庭用品 |
| --- | --- | --- | --- | --- |
| 阿馬蘭特 | 約6000人 | 48間 | 27家 | 51家 |
| 哈靈頓 | 約8000人 | 11間 | 12家 | 12家 |

# 3.地區設施的維修

## 阿馬蘭特及哈靈頓

在葡萄牙北部，以香檳酒產地聞名的米紐地區南方，有一個稱爲阿馬蘭特的小鎮。

根據歐洲旅遊指南的簡介，這個小鎮的人口約不到五千人，但若是到該地的鎮公所去詢問的話，他們都表示人口約有六千人。但是無論如何，這都算是一個鄉下小鎮。但這個小鎮周圍的農村地區，有很多外來的客人，所以也算是十分熱鬧的。

在鎮裡的往來行人的多寡程度，由酒吧（可以喝酒，喝飲料以及吃點心的地方）的數量就可以得知了。如果繞行小鎮一圈數數看的話，可以發現全部竟然有48家。其他的乾果店、麵包店、肉店等食品店有有27家，雜貨店、化妝品百貨店、家俱店、及電器用品電，等雜貨及家庭用品等，全部有51家。對葡萄牙的這個小鎮阿馬蘭特的人口規模而言，這些商店的數量究有多驚人，我們可以和歐洲其他人口規模相似的其他村鎮做比較。

我們就拿英國，蘇格蘭首都愛丁堡東南方，約三十公里處的小鎮哈靈頓來做比較。

哈靈頓是蘇格蘭低地肥沃地帶的一個農業區，人口大約有8,000人左右，是一個比阿馬蘭特稍大的村鎮。以單純的觀點來考量的話，哈靈頓的商家，應該會比阿馬蘭特多許多。但是，在哈靈頓的小酒館（與葡萄牙酒吧相類似的規模）和吧台等數量總共有11家，和阿馬蘭特的48家的規模，實在是無法相比。同樣地，賣食品的商店共有12家，雜貨店、家庭用品店總共也是只有12家，所以仍是比阿馬蘭特少很多。（請參考表四）阿馬蘭特和哈靈頓商家數量的不同，有一個原因是每家商店的規模並不相同（哈靈頓每家商店的面積較大），但是這並不足以說明這個壓倒性的數量。阿馬蘭特會有壓倒性數量的原因，當然也是前來的客人數量較多，需求量較大之故。但因爲阿馬蘭特的人口並不多，所以前來的多半是外來的觀光客（相反地，哈靈頓就沒什麼外來的客人）。也就是說，葡萄牙和英國村鎮最大的不同，是在於村鎮周圍農村地區的有無。

相對於葡萄牙的農業就業人口，占全國就業人口的18.7%（1987年），而英國在同年還是不超過2.2%。由此可知，在農業比占經濟全體較少的先進國家，因農村地帶的人口少，也顯示出農村地區人口爲基礎的商業發展也越脆弱。反過來說，因經濟發展而致使農業比重低下的情形而言，要維持村鎮內居民的數量，就要取代從事商業的人口，進而增加住宅區裡的人口。

## 人口規模和設施的種類、數量

對於現代人而言，要經營一個快適，舒服的生活，必須要有一個具備各種地域設施等的便利住宅環境，例如：日常購物的商業設施、學校、醫療設備、文化及娛樂設施等。但是，在人口較少的聚落，各種地方設施的數量及種類也較少。而且如前所述，一旦成爲先進國家之後，因農業人口的減少，即使是規模相同的村鎮，也是越先進的國家農業人口越少，而商業設備的數量也越少。

在表五中，是筆者由之前的阿馬蘭特和哈靈頓爲始，進而在歐洲幾個各小村落實際地往來察看之後，所做的一個統計。但是因爲在調查時，也受到步行範圍的限制，特別是設在周圍邊緣帶的教育設施或是休閒娛樂設施，或許多少會有些遺漏。尤其是葡萄牙和西班牙的調查資料，因爲公事包被竊，所以關於葡萄牙和西班牙的數字，是後來憑記憶記錄的，因此不能確保其精確度。但是可以確認的是，因村鎮人口規模不同，而在區域性的設施設備及水準，也有明顯的差異，還有如前所述地，因經濟發展而致使農業人口比例減少時，會使得商業設施設置的種類及數量也相對地比人口規模小。例外的是南斯拉夫的奧米休，因爲奧米休這個小鎮是位在，面對亞得里亞海的大都市——斯普里特郊外約二十公里的地方。這個小鎮不但面海而且還有岩山爲屏，所以不算在農業帶的經濟圈內，再加上南斯拉夫是社會主義國家，無法自由進行商業交易，也是其原因之一。

另外，如表六（下頁）中所謂，根據調查及視察，參考各都市計劃的報告和規劃者（主要來自英國），所做的因應各村鎮人口規模、一般土地的利用及地區設施之概要摘錄。就如表格中所示，對歐洲各先進國家來說，村鎮等地方性的設施，會因村鎮本身的種類及數量等各有其限定。爲什麼，在設施種類和數量受到限制的歐洲各先

和民宅的村鎮。

小村鎮要和觀光共存的一項要領，就是要能有效地控制遊客的人數。如果過多的觀光客前來，不僅只會影響到當地居民的生活，歷史性的文化史蹟或建築，也會因大量遊客的手垢及往來而造成的底盤摩損，濕氣等而損及建物或結構體。但是，和大城市（例如：巴黎的聖母院等）相較之下，這類問題在各村鎮間並不那麼嚴重。

而以上提及的控制觀光客的人數，就可以用停車場的數量及位置達到控制的效果。例如：在附近的停車場的停車位數量，大約就是平時一般觀光客的使用量；而在觀光旺季，觀光客較多時，車子不得不停放在較遙遠的停車場，如此一來，就必須得有一段不短的步行距離，而觀光客因爲這樣，也不致於會大批湧入了。在觀光淡季時，因鄰近處就有停車場，相當便利，所以客源也不致於減少，而且在旺季時，雖然停車場不夠，但也不致於全無停車位，而讓客人敗興而歸。還有在小村鎮中，要建出這樣的停車場也並不困難的。在平時，幾人同行結伴而來時，村鎮內精緻的餐廳，咖啡屋，時裝店等等，對於創造村鎮的便利及舒適性，其實有很大的助益。

另一方面，因觀光客源增加而大舉興建飯店，整修旅館時，必須要注意以不損及街道的景觀和不破壞整體美感爲原則。對村鎮來說，比較方便的是可以放置露營車及大蓬車的設備，這樣就可以因應觀光食淡季和旺季來置放處理，所以也是十分方便的作法。這些露營車和大蓬車可以在淡季時收拾起來，之後也無需人手來照管，也就不需要再續雇旺季時的勞工了。

在維持美麗街景和發展觀光的同時，有一個無可避免，而且會隨著這些發展越來越嚴重的問題。那就是房地產價格的提昇。舉例而言，英國湖水區的房子，近年來因大家購置別墅的需求增高，所以在1970年，房地產價格比英國平均房價底約百分之二十至三十的湖水地區，到了1980年時，該地區的房價已經揚昇，並較平均房價高出約百分之二十到三十。這樣的情形雖然對售屋者較爲有利，但是對於欲購屋者（因房租調高）及租賃而居者來說實屬不利，但是住在該地區的居民，卻逐漸被購置別墅者（或租用者）及退休養老者所取代。不過假設原當地居民將房子出售而搬離，至少對原先擁有房屋者而言，可以用較高的價格售出也未嘗不是件好事。所以因觀光所引起的不動產價格升高，對社會至少應該有經濟上

在人口稀少的英國鄉村，郵局通常由雜貨店兼任。巴頓，布蘭德斯多克。

學校是除了郵局之外，最普遍的公共設施。學校的建築往往都是十分大規模，而且常和周圍的街道景觀產生不協調的設計。但在這幀校和周圍的街道景觀，卻是少見地和諧美麗的調和。法國，奧特瓦爾。

進國家的村鎮，仍能保有並維持其舒適及便利的生活呢？

## 地方設施擴大範圍地維修及共同使用

隨著從事農業的人口減少，在地方村鎮上設置並維持都市生活必須的地方設施之中，要維持設置商業設施的困難自不待言，但是公共設施方面竟也十分困難。公共設施中最典型的例子，就醫療設備來說好了，如表七面所記的資料中，雖然資料不是最新的，但是這是在1955年法國，醫療專業區分之中，一個專科醫生平均的受雇人口調查。

因爲醫療技術的進步，所以爲了提升醫療效果地，也將各專科醫生以其專長而細分出來；表七中，十分明顯地可以看出來，對專業醫師的需求量很低。在一千五百人之中才需要一名一般的內科醫生，所以在人口一千人以下的村落裡，周圍村落加起來，共同只需要一名內科醫師。同樣的，一般內科、外科再加上泌尿科，耳鼻喉科，眼科，婦產科，放射科等的綜合醫院，爲了設置在小村鎮中，所以就必須和鄰近的村鎮共同擁有使用。而兼具心臟科，胃腸科，放射線治療科，專門兒科等稍具水準的大型綜合醫院，更甚者像是包含神經外科，肺臟外科，或是癌症的專門治療等，具大學附設醫院水準的這種以廣大民眾爲主要對象的綜合醫院，就必須設在較具規模的都市核心地區了。也就是說，當在小村鎮中，需要現代化較高度精細醫療時，就必須要送往周圍較大的城鎮都市核心區的綜合醫院。

同樣的情形，有很多文化及商業設施也是依循這樣的模式。想要欣賞歌劇或是音樂會時，還是必須前往擁有音樂廳及歌劇院的鄰近中型都市去。就東西方面來說，一般的食用品及日用雜貨，在自家的村鎮就是可以買到了，但是若是想要購買較高級的東西及傢俱等的時候，還是非到較大的城鎮去不可。也就是說，在小村鎮中，要維持這些現代的都市生活設施，除只需土地不用建設、維修和營運費用的公園，運動場等設施之外，人口較少的村鎮，仍是無法負擔過大的地方設施，或者可以說是太困難而不可能實現。因此，爲了創造村鎮的舒適性，在考量整備廣大區域的共同利用性時，如何利用附近村鎮和市鎮中

● 表五　各村鎮的施備狀況一覽表

| 村鎮名稱 | 奧米休 | 馬爾波 | 扁比杜斯 | 卡什特格 | 阿馬蘭特 | 巴塞盧斯 | 德爾馬洛 | 卡西亞諾 | 聖塔傑洛 | 馬特爾 | 霍克斯菲德 | 洽壹頓 |
|---|---|---|---|---|---|---|---|---|---|---|---|---|
| 人口 | 8000 | 309 | 829 | 2558 | 6000 | 7500 | 3884 | 1000 | 16582 | 1441 | 500 | 8000 |
| 國別 | 南斯拉夫 | 葡萄牙 | 葡萄牙 | 葡萄牙 | 葡萄牙 | 葡萄牙 | 西班牙 | 義大利 | 義大利 | 義大利 | 英國 | 英國 |
| 農業就業人口比% | 25.0 | 18.7 | 18.7 | 18.7 | 18.7 | 18.7 | 12.4 | 8.3 | 8.3 | 8.3 | 2.2 | 2.2 |
| **公共設施** | | | | | | | | | | | | |
| 郵局 | 1 | 1 | — | 1 | 1 | 1 | — | 1 | 1 | 1 | 1 | 1 |
| 警察局 | 1 | — | — | 1 | 1 | — | 1 | 1 | 1 | 1 | — | 1 |
| 消防隊 | 1 | — | 1 | — | 1 | — | — | — | 1 | 1 | — | 1 |
| 觀光導覽處 | 1 | 1 | 1 | 2 | — | 1 | — | — | 1 | 1 | — | — |
| 其他 | 1 | — | — | — | 1 | — | 1 | — | 3 | — | 1 | 3 |
| 公共廁所 | — | 3 | 2 | 1 | — | 1 | — | 1 | 1 | 2 | — | 1 |
| **教育設施** | | | | | | | | | | | | |
| 幼稚園 | — | — | — | — | 1 | — | — | — | 1 | 1 | — | 2 |
| 小學 | 2 | 1 | 1 | 1 | 1 | 1 | 1 | 1 | 1 | 1 | 1 | 2 |
| 中學 | — | — | — | — | 1 | — | — | — | 1 | — | — | 2 |
| **醫療設施** | | | | | | | | | | | | |
| 綜合醫院 | — | — | — | — | — | 1 | — | — | 1 | — | — | — |
| 療養院 | — | — | — | — | — | — | — | — | 1 | — | — | — |
| 診療所 | 2 | — | — | 1 | — | — | — | — | 1 | — | — | 1 |
| 醫院兒 | — | — | — | — | — | 7 | — | — | 1 | 1 | — | 1 |
| 牙科診所 | — | — | — | — | — | 3 | — | — | 1 | — | — | 1 |
| **飲食及住宿** | | | | | | | | | | | | |
| 餐廳 | 10 | 2 | 5 | 2 | 5 | 4 | 8 | 2 | 11 | 2 | — | 3 |
| 酒吧及其他 | 28 | 2 | 6 | 5 | 48 | 61 | 6 | 2 | 21 | 3 | 4 | 11 |
| 飯店 | 1 | 1 | 2 | 1 | — | — | 3 | — | — | 1 | 2 | — |
| 民宿 | 15 | 2 | 2 | 10 | 3 | — | 17 | 1 | 2 | 4 | 6 | 8 |
| 露營車專外宿 | 2 | — | — | — | — | — | 2 | — | — | 1 | — | — |
| **服務業** | | | | | | | | | | | | |
| 汽車修理廠 | 1 | — | — | 2 | 4 | 9 | — | — | 2 | — | — | — |
| 其他修理廠 | 1 | — | — | — | 3 | — | — | — | 2 | — | — | 2 |
| 洗衣店 | — | — | — | — | 2 | 3 | — | — | — | 1 | — | — |
| 美容理容院 | 5 | 1 | — | 1 | 6 | 15 | — | — | 7 | 4 | — | 4 |
| 銀行 | 2 | 1 | — | 1 | 2 | 4 | 5 | — | 1 | — | 2 | 4 |
| 其他金融機構 | — | — | — | — | — | — | — | — | — | — | — | 5 |
| 保險機構 | — | — | — | — | 1 | 2 | — | — | 3 | 1 | — | 1 |
| 不動產機構 | — | — | — | 1 | 2 | 2 | — | — | 4 | 1 | — | 7 |
| 專門的服務機構 | — | — | — | — | 4 | 4 | — | — | 6 | — | — | 4 |
| 其他 | 6 | — | — | — | 8 | 8 | — | — | 6 | 1 | — | 2 |
| **小型販賣店（食品類）** | | | | | | | | | | | | |
| 一般食品 | 3 | 1 | 2 | 5 | 10 | 15 | 1 | 2 | 11 | 2 | — | 5 |
| 麵包店 | 3 | — | 3 | 5 | 8 | — | — | — | 1 | 1 | — | 2 |
| 肉店 | 1 | — | 4 | 6 | 10 | — | — | 1 | 7 | 2 | 1 | 2 |
| 魚店 | 1 | — | — | — | 1 | — | — | — | 1 | — | — | — |
| 蔬果店 | 1 | — | — | — | 1 | 2 | — | — | 1 | — | — | — |
| 其他 | — | — | — | — | 4 | 10 | — | — | 5 | 2 | — | 1 |
| 報紙，點心，香煙 | 7 | — | — | — | 4 | 8 | — | 1 | 6 | — | — | 4 |
| 藥局 | 1 | — | — | 1 | 6 | 9 | 1 | 1 | 4 | 2 | 1 | 3 |
| **衣料服飾店** | | | | | | | | | | | | |
| 一般衣料服飾 | — | — | — | — | 3 | 15 | — | — | 1 | — | — | 1 |
| 男裝店 | — | — | — | — | 7 | 8 | — | — | 2 | 1 | — | 2 |
| 女裝店 | 5 | — | — | — | 15 | 20 | — | 3 | 6 | 2 | — | 6 |
| 其他 | 1 | — | — | — | 3 | 5 | — | — | 1 | — | — | 3 |
| 鞋及皮飾店 | 4 | — | — | — | 10 | 15 | — | — | 3 | 1 | — | 3 |
| 衣料 | — | — | — | — | 3 | 2 | — | — | 2 | — | — | 3 |
| **（家庭用品及五金）** | | | | | | | | | | | | |
| 家具 | — | — | — | 1 | 5 | 12 | — | — | 2 | — | — | 2 |
| 電器用品 | 1 | — | — | 2 | 12 | 12 | — | — | 2 | — | — | 3 |
| 工藝品及土產品 | 14 | 1 | 8 | 2 | 22 | 33 | 13 | — | 25 | 4 | 5 | 1 |
| 骨董品 | — | — | — | — | — | — | — | 2 | 1 | 2 | — | 1 |
| 其他 | 2 | — | — | 3 | 12 | 15 | — | — | 7 | 1 | — | 3 |
| 汽車 | 1 | — | — | — | 3 | 8 | — | — | — | 2 | — | 3 |
| 加油站 | 1 | — | — | — | 2 | 4 | — | 1 | 4 | 1 | — | 3 |
| 書店及文具店 | — | — | — | — | 2 | 4 | — | — | 3 | 1 | — | 1 |
| 其他 | 7 | — | — | — | 13 | 25 | — | — | 7 | — | — | 12 |
| **娛樂及文化事業** | | | | | | | | | | | | |
| 圖書館 | 1 | — | — | — | 1 | 1 | — | — | 1 | — | — | 2 |
| 博物館及美術館 | 2 | 1 | — | 1 | — | — | 4 | — | 1 | — | — | 1 |
| 劇院及音樂廳 | — | — | — | — | — | 1 | — | — | — | — | — | 1 |
| 電影院 | 2 | — | — | — | 2 | 1 | — | — | 1 | — | — | 1 |
| **（運動設施）** | | | | | | | | | | | | |
| 網球練習場 | — | — | — | — | — | — | — | — | — | — | — | 1 |
| 運動場 | 1 | — | — | — | 1 | — | — | — | 1 | — | — | 2 |
| 室內游泳池 | — | — | — | — | — | — | — | — | — | — | — | 1 |
| 室外游泳池 | 1 | — | — | — | — | — | — | — | 1 | — | — | 1 |
| 體育館 | — | — | — | — | 1 | — | — | — | — | — | — | — |
| 高爾夫球場 | — | — | — | — | — | — | — | — | — | — | — | 1 |
| 散步步道 | 1 | — | — | 1 | 1 | 1 | — | — | 1 | — | — | 2 |
| 公園 | 1 | — | — | 1 | 1 | 2 | — | — | 1 | — | — | 1 |
| 兒童公園 | — | — | — | 1 | 2 | 2 | — | — | 1 | — | — | 1 |
| 其他的遊戲設施 | 1 | — | — | 2 | 5 | 4 | — | — | — | — | — | 1 |

在葡萄牙東南部,波爾圖阿雷格特附近的蒙福特發現的小醫院。在葡萄牙的小村鎮裡,這類的小規模的醫院很多。但是在英國等地區性大醫院較發達的國家,村鎮裡這種小醫院就不多了。

這是蘇格蘭爲了美化田園景觀所進行的一項活動。最近的活動點在於要如何營造都市周圍的田園景觀,也就是要創造富有景觀性的袖珍型都市。照片中是在展示郊外用的看板。在蘇格蘭的帕斯附近,雷德戈頓。

● 表六 歐洲各先進國家的村鎮大小·規模·土地利用·區域設施

| 型式 | 主要的經濟基礎 | 人口規模 | 經濟性的影響範圍 | 村鎮建築物的主要用途 | 區域設施 |
|---|---|---|---|---|---|
| A1 | 農業 | 100未滿 | 可以忽視之(3km以內) | 農舍的聚落 | 無(因情況而異,有時郵局會兼任小雜貨店1＋教堂1) |
| A2 | 農業·(商業) | 100—1,000 | 非常小(3-5km) | 農家和某些商店組成的村落 | 有時郵局會兼任小雜貨店1,汽車服務中心1教堂1(因時而異,旅館有時也會兼營咖啡廳(酒吧)1小學1銀行1醫院1) |
| A3 | 農業·商業 | 1,000—10,000 | 小(含村莊)(5-10km) | 中心區爲商店和公共設施,周圍是住宅區的小鎮 | 專門店1-40,中學1圖書館1公共建設(因時而異,可能是活動中心或綜合醫院) |
| I2 | 農業·工業 | 100—1,000 | 非常小(附近有工廠)(3-5km) | 農家和普通住宅集合而成的村落 | A2的設施,加上食品行1酒店(酒吧)1另外,還有咖啡廳或餐廳1。 |
| I3 | 工業·商業 | 1,000—10,000 | 小(5-10km) | 有商店,工廠以及住宅的村鎮 | 和A3一樣,(會因時而有工業區加入)。 |
| I4 | 工業·商業 | 10,000—100,000 | 中等(10-20km) | 有商店,辦公室,工廠及住宅的中型村鎮 | 和I3的設施相同,但數量及種類較多,還有旅館1-5,文化或是運動設施(例如:電影院,美術館等設施)。 |
| R2 | 休閒·觀光 | 100—1,000 | 非常小(除了休閒和觀光客之外) | 主要爲商店和別墅的村鎮 | A2的設施再加上民宿,藝品店,咖啡館或是餐廳等。 |
| R3 | 休閒·觀光 | 1,000—10,000 | 小(除了休閒和觀光客之外) | 主要爲商店和別墅的村鎮還有旅館所構成的小村莊, | R2和A3的情況再加上旅館1-10(含賓館等),附近還有露營車等設備。 |
| C1 | 住宅 | 100未滿 | 可以忽視之 | 爲了通勤人員而將農舍改造成的住宅聚落 | 和A1相同。 |
| C2 | 住宅 | 100—1,000 | 非常小 | 爲了通勤人員而將農舍改造成的住宅,還有一些商家組成的村落 | 和I2相同。 |
| C3 | 住宅 | 1,000—10,000 | 小 | 爲了通勤人員而將農舍改造成的住宅聚落 | 和I3相同。 |

● 表七 醫療之專門分類別中,雇用一個專任醫師的平均人口

| 專門科別之分野 | 專門醫療分類中,雇用一個專任醫師的平均人口 |
|---|---|
| 一般內科 | 1,500 |
| 一般外科 | 20,000 |
| 泌尿科、耳鼻喉科、眼科、婦科、放射線科 | 40,000 |
| 心臟科、胃腸科、放射線治療、專門小兒科 | 100,000 |
| 神經外科、肺外科、癌症的專門診療 | 2～400,000 |

的商業設施及公共設施,是十分具有重大意義,而且值得深思的。

支援小村鎮便捷及舒適性的小型市鎮

在考慮整合地方設備以創造村鎮快適便捷的同時,對地方上而言,以含蓋中型核心都市及村鎮的地域觀念來思考,是非常重要的。也就是說,住著數萬人的市鎮或是有著二十至四十萬人的中型核心都市的都市計劃,是會對周邊的小村鎮造成重大影響的。

大市鎮或中型核心都市的都市計劃,應該是要以在較爲廣闊的區域概念爲考量基礎,例如在整合道路及停車問題時,也必須充分考慮到市鎮外的居民來時的停車問題及道路問題。因爲對先進國家來說,較邊僻地區的居民,所依靠的交通工具,就是汽車。所以如前「1.對於汽車社會應對」中所提及的,在汽車社會中速度的重要,道

路沿線帶狀開發的控制,以及應該在靠近市區中心處設置公共停車場等。如果能將這些條件視爲整合時的要點,則外環地區的村鎮居民就可以利用核心區的商業設施及公共設備,而且不僅於此,這也是維持市鎮和核心都市繁榮所必須做的努力。或者我們也可以說,要維持附近大市鎮和核心都市的繁榮,創造村鎮的安適及便利是不可或缺的條件。

這也表示了一件事,就是村鎮的人民,可以用汽車爲工具,來享受市鎮及核心都市的商業設施和公共設備,而且也不用因爲要享受這些設施,而居住在大城市之中。所以我們可以得到一項結論,就是爲了維持都市周邊地區的田園地帶,防止都市人口過密,並且享受汽車的便利快速,中型核心都市的規模就不可以過大(自市街外緣地區至中心區,約十分鐘左右的車程爲最適),以

適度的大小爲宜。

事實上在英國,爲了防止大都市的人口過密,爲了維持田園環境,而在田園外圍開始建設含有辦公室及工廠的新市鎮。現在已經無法再興建新市鎮了,但是不僅只是大都市,就連中型核心都市,都必須抑止其街市的擴張。而新的住宅區,不再是建在市鎮或中型核心都市的外圍,而是建在小村鎮之中了。爲了提高村鎮的舒適和便捷,中型都市也都以不再擴大爲原則。不光是這樣,連工廠及辦公室等工作場所,也開始逐漸搬進村鎮的附近了。這就是歐洲各先進國家所謂的「田園革命」。

# 設計上的解說

## 起源

在歐洲，擁有各種歷史背景的小村鎮，往往能保有自身以獨特的地建築及生活形式為原則的傳統街道景觀，又具彈性地能適應現代生活，是以其豐富的獨特性和充滿舒適性的魅力街市組合而成的。

但是在現在的歐洲各村鎮，究竟是如何地保持村鎮的傳統街景，並創造出具魅力的舒適環境的呢？在這裡，就這個創造舒適空間的考量，轉移到現實時，我們就在「計劃」的部份裡解說。接著，下面就對於在小村鎮中進行計劃之節，儘可能做具體的印象做介紹。

在英國蘇格蘭羅吉安地區，對東羅吉安（哈靈頓市東方的小村鎮，總人口約八萬人的農村）地區，為了推行這計劃制度及運用方針，也為了讓當地居民明白這計劃的目的，還發行了宣導手冊，以倡導說明之。並且同樣進行這種計劃的還有含英格蘭地區的英國全國，及歐洲其他國家的計劃，也就是類似這種形態。

也就是說，不管是在哪個國家，歐洲的各村鎮都是以街道的美觀及舒適性，做為計劃的主幹，而創造保存這些街景的。當然，計劃的制度及實際上之運用，會因各國家、地區和體制不同，而各有差異。但是唯一的共通點是在實行時，各國都用一日本絕對望塵莫及的嚴格態度來執行的。例如因為是自己的土地，就隨意改建等等的行為，是絕對不容許的。即使是讓大家覺得已經損失掉許多文化資產的義大利，若是計劃未得許可以前擅自改建，則不但要自費將建築物回復原狀之外，還必須負擔一筆可能要房子才付得起的罰金。

## 由誰來執行計劃呢？

在東羅吉安地區，所有的計劃業務，幾乎都沒有委託外部單位，而由地方政府的設計部門所屬的，受過設計專門教育的十來名職員所為的（部長是具有博士學位的）。這些職員通常會為構成地方議會的「計劃及開發委員會」的議員助言，而由被選中的議員來進行計劃最後的決定。設計和計劃，最好不要借外人之手，是歐洲各國一致的共通態度。

## 計劃的兩大機能

1.制定計劃：寫下將來利用土地和建物使用

（哪的道路需要拓建，哪裡要新建住宅等等）的計劃方針。

2.計劃的許可：對於具體的土地及建物，決定是否應該認同，並同意如人們所提之要求般改變（也就是核發設計計劃的許可）。

在各國的計劃之中，1.詳細地執行的制定計劃，2.在取得計劃許可時，計劃者不再有斟酌變更的空間；另外是1.不那麼詳盡地制定計劃，但2.的許可取得時，設計者還可以有較多的轉寰性，是相當具變化性的。在歐陸諸國的設計，通常一般是採用前者，英國則是採行後者。只是在這裡有一個重要的事實，就是無論將來設計者的計劃多麼地週詳，如果不能確實地發揮設計的功能，則計劃的實現也是十分不可期待的。以這個觀點來考量的話，在維持及改善小村鎮的街市景觀時，要將計劃落實在生活中，發揮其威力，應該可以說重點是在於計劃許可，而不是在於制定計劃。

## 制定計劃

一般而言，不管在哪個國家，會因區域大小或目的不同，而有幾種不同的計劃。英國，像是在東羅吉安這樣較小的行政地區進行的是「地方性的計劃」，還有以廣大地區（除了東羅吉安之外，還包括中羅吉安、西羅吉安等）為對象的「概略性的計劃」等兩種形態。

### 概略性計劃

以廣大範圍為對象的結構性計劃，是針對未來在創造地區的目標和戰略之上，確保這些將來對交通、住宅、產業等是為必要性的土地，做戰略性的計劃＜方針＞及＜提案＞。這種＜方針＞和＜提案＞，是以模式圖案和文章來表示計劃的整體。所謂概略性計劃的主要目的，是在於計劃核准之後，為了規劃以為基準的地方性計劃，而以更廣闊的視野，所做的概略性架構。

例如羅吉安地區的概略性計劃中，針對東羅吉安做以下的策略式提示。

1.集中東羅吉安地區（和哈靈頓相連接）的開發，以抑止其他六個區域的發展。

2.支援經濟開發。

3.保護農地，提高既有村落城鎮的舒適和便捷性。

4.促進自然和街道的保存活動、休閒娛樂、觀光等事業。

但是，這種概略性的計劃並沒有詳細地記載在地圖上，所以並不能回答出一般人想知道的，現在住著的馬路或工作的地方，將來會成為什麼樣子。但是若是以小地區為主的地方性計劃，就能回答這些大家心中的問題，因此，地方性計劃和概略性計劃，就是以這種相輔相成的形態而存在的。

### 地方性計劃

地方性計劃，是將概略性計劃中所提示的＜方針＞及＜提案＞，付諸行動的實際動作。所以具體地將方針及提案，實際運用於何處等等，詳細地標示於地圖之上，將其整體表現出來。地方性計劃，是把計劃一點滴地在地圖上揭示出來，例如將經過許可，可以興建新的住宅，辦公室，工廠等的地方標示在地圖上，更進一步地將禁止汽車駛入或是拆除妨礙視線的招牌和建築物，改善景觀，並且標示出馬路，地點等等。另外，地方性計劃中所提出來的方針及提案，也可以說是決定計劃許可（關於在何種情形需要核准許可，將於敘述於後文中）時的基準。因此，對一般民眾而言，地方性的計劃比概略性計劃更具有實質意義。

在東羅吉安地區，在約十年前開始，就將地區劃分成七個小區域，並且各別制定詳細的計劃，定期進行修訂。也就是說，相對於在日本，都市計劃對象僅只在有限的街市或是新市鎮裡，並且敷衍著之的情形相較之下，在英國，無論是多小的村鎮或村落，還包含田園地區，全國的各角落都已規劃在計劃網中，根據各地區的特性制定將來的具體計劃，並且持續進行。

### 鄧巴新制

進行地方性計劃的七個規劃區中，有一個做為公開高爾夫球賽而聞名的地方，含鄧巴市鎮之地區。鄧巴的市鎮，造成建設風潮，等興建告一段落之後，隨著經濟的沈寂，中心區的街道和建築物的荒廢是隨處可見的。建設風潮是一時性的，如果不進行長期性的投資，簡單地建設，會給市鎮發展帶來不良的影響，這個鄧巴的例子，也不是例外。

在鄧巴市區中心，被指定為保存地區，除了面向中央大馬路的建築物，大多數（約70間強）是指定建築之外，大部份的建築物也對這歷史性的

在鄧巴中央大馬路上，掛有招牌鄧巴新制的事務局。

鄧巴新制開始進行已經五年了，雖然計劃也未必能如預定般進行。而是在和過去相較之下，鎮上的環境改善了，現在村鎮上已經可以感受到鮮明愉快的氣氛了。鄧巴的中央大馬路上。

街景做了貢獻（關於保地區指定建築物的計劃，詳如後述）。所謂鄧巴新（DUNBAR INITIATIVE），是因為了要再度帶動環境和經濟，以市中心的保存區為對象，加以具體地改善，詳細來討的行動及過程。將面對街市的建築物，都以縱向剖面圖來標示，再加以計劃色彩，內院中荒蕪了的建築物，則以照片中荒蕪了的建築物，則以照片佐證，來製定這些建築物的改建方案，詳細地調查既有土地的利用，土地的開發，或是和周圍建築協調的新設計建築，還有重要幹道的停車場計劃等，大部份都是以圖示或草圖，來檢討提案，再將施工費用及估算表等表示出來。我們在拜訪東羅吉安的設計部長（1988年秋天）時，正好遇上這個鄧巴新制（DUNBAR INITIATIVE）的案例，為了實踐這個構想，籌措公共資金，抱持著關心來發展，是今後的重要課題。

## 設計摘要（PLANING BRIEF）

「設計摘要」是為了能詳細地指示出，在地方性計劃中，關於在開發預定地（例如損了的連續房舍之建築用地等）上的發展時，應該進行什麼樣的開發等。「設計摘要」的作用在於，它可以供和街景調和的建築物設計等具體的，且在地方性計劃中沒有標示的具體方針。因此，在為進行修築而檢討建築場地時，不僅只是地方性計劃，還必需要事前調查好是否有「設計摘要」的規範。例如在鄧巴新制（DUNBAR INITIATIVE）中，設計摘要（PLANING BRIEF）就發揮了作用了。

## 變更用途及改建

關於土地或建築物用途變更時，也希望大家在事前先向所屬相關單位查詢清楚。例如：以往的農莊或是教堂要改為住宅時，或是大房子要分為幾個公寓單位時，是一定需要設計許可的。另外，像是過去取得設計許可，將倉庫改為室內體育館的例子，若是想再將室內體育館改為倉庫時，也還是必須再取得設計許可的。

但像是維護建築物外觀的一般作業，例如：粉刷油漆，交換窗櫺等通常是不需要設計許可的。另外房屋內部的改裝等，通常也不會設計許可。但是若是保存對象的指定建築物，或是保存區內的建築，即使是公寓內的內部改裝等，也是必須申請設計許可的。關於指定建築物或是保存區的設計許可等，將在後面章節裡詳述。

若是因建築物的新建、增加改建或是變更建築物用途，伴隨而產生的改建時，必須要取得不同於設計許可的建築許可（BUINDINF WARRANT）。

建築許可以目的，是為了確保建築物本體的構造性和環境安全性。在所有即使只是稍作修改的時候，也幾乎都需要取得和設計許可不同的建築許可。

## 設計申請的程序

設計許可的申請通常需要花八週（因狀況不同，可能會花上更多時間）以上的時間，所以及早申請是關鍵。申請的步驟是：

1.首先確認設計許可是否為必要。

2.提出申請前，先和政府的設計部門的開發規定負責人（設計者）商量。因為最後判定設計許可的是地方議會的「設計及開發委員會」，但是可以請教設計人員並尋求協助，以判斷許可申請是否可以通過，或是申請可能不易通過時，也可以請他們就問題點加以說明。另外，也可以請教他們申請計劃要如何修正才可以。

或者是有時候也會建議大家先申請「輪廓草圖（OUTLINE）許可」。這是在考量委員會是否會核准的微妙狀況時，可以先進行的全盤性計劃許可的申請，如果這個許可獲得了同意，再具體地提出詳細申請；而且這個方法還可以省下委託建築的費用。此外，在一些保存區中，「輪廓草圖（OUTLINE）許可」也必須附加設計概略圖。

3.確認申請書之記載，要能正確無遺地提出。在很多時候，可能必須藉助於建築師或是不動產鑑定師等。而且申請時，除了要支付申請費用之外，也必須要提出兩種證明的文件。一種是所有權的證明，若非本人所擁有之土地或建築時，必須要有已告知所有人之證明；另一種是申請預告及通知的證明，就是通知附近居民，有關設計申請的告知證明。若是高度超過20M以上的建築，亦或是興建競技場、游泳池、電影院等遊樂設施，或者是動物園、動物商店、屠宰場等，以及公廁，地下水處理場、廢棄場、礦坑、墓地等，被稱之為惡鄰的開發內容，在申請時，都必須在地方性報紙的廣告欄上刊登，並且清楚地刊載地址及詳細狀況。

4.確認申請書的受理通知。

5.等待委員的決定。通常需要八週左右，但是有時也會快一點。在這段期間，設計者會詳讀申請書，看是否符合地方性計劃（地方性計劃不適用時，則改採概略性計劃）或是計設概要的＜方針＞，而且還要和相關的機構討論道路、水管等問題，從計劃的所有層面來作檢討。有時候為了決定與否「設計及開發委員會」還會親自到現場

進行考察。這時便會通知申請人前往，除申請人之外，也允許申請人之外的第三人同行。

6.許可：也分成無條件許可，附帶條件之許可，不許可三種情形。獲得許可時，在得到建築許可的期限（通常是五年）內，隨時都可以開始動工。

## 不服申請之提出

若是對於設計不許可或是對於許可　附帶條件不服時，可以在自通知日起的六個月內（屋外廣告的情形是兩個月，植栽則是28天之內），對國家（蘇格蘭的話，就是蘇格蘭的相關負責大臣，英格蘭就是環境大臣）提出上訴（APPEAL）。提出時，必須在規定的紙張格式上記入必要事項以及提出上訴的證據。在提出上訴的處理方式上，通常是召開公聽會，再由來檢查官作出許可（通常會有附帶條件）或不許可之結果。不只是不許可的情形，就是對於許可的附加條件提出上訴時，即使是有證據的上訴，也未必會通過。或者是條件過於嚴苛時，有可能也會有取消許可的情形發生。而且，因為從申請到結果之判定，可能要花一年以上的時間，所以在提出不服申請時，請先謹慎地考慮。

# 對申請案提出意見時

## 設計申請是被誰通知的呢？

因為核發給申請人設計許可，所以那些擔心自己的利益受到侵犯的人，無論是誰，都可以對申請案提出反對或是意見。就是因為這點，所以在申請設計許可時，必須對所有相關人進行通知。在對非本人持有土地或建築物進行設計申請時，設計許可也可能會對土地持有人造成利益上的損害。因此申請者在進行申請之前，有義務且必須先知會土地或建築物所有人有關計劃的內容等。另外，對相鄰土地或建物持有人亦須進行通知。但是在道路的對側時，可以不進行知會動作。

小酒店（酒吧），速食店等營業場所，以及先前所提及歸為惡鄰之開發，還有指定建築物（於後文詳述之）之相關的設計申請，必須進行大範圍的通知，通常是在地方報紙上刊登廣告，以公告現有土地及建物之計劃內容。

除了公開的公告之外，很多人會由地方新聞記事中得知，或是因大家的口耳相傳而得知。無論

是誰，用何種方法得知，若是想更進一步得到詳細資料時，都可以從由地方政府保管的申請櫃台中借閱。在底稿處，除了申請書之外，還會附有圖面及概略草圖。若有難懂之處，也可以請相關人員說明之。

## 提出意見的時限

提出意見的期限，通常是二至三週，一過了這段時間，只要計劃的申請通過決定之後，所有的反對或是意見都不會被考慮採行的。因此若反對或有意見要提出時，把握時限，迅速應對是最重要的。

## 提出意見的方法

在對申請計劃設計時，不只是直接有利害關係的當事者，無論是誰都可以提出反對（或是贊成）意見的。提出意見的方法，就是寫信給負責的辦事員或是長官，清楚地寫下，對哪個申請有什麼意見，並且將反對（或是贊成）的理由寫下來。

若是認為設計許可必須加附條件時，請也詳細敘述需要何種條件，以及為何需要這種條件。在「計劃及開發委員會」對該計劃進行設計申請審查時，這些意見或是反對的原因，都是在決策時，必須考慮的意見。

可是在英國的設計計劃的制度下，這有時候會無關乎反對地認定設計申請時，不只是土地持有人，或是法定權利者（依據查判定等）也不能阻止其開發。

## 關於指定建築物的設計

在英國，以設計的保存對象，嚴格地篩選登錄於「在建築或歷史上特別具有價值的建築物一覽表」的數量相當多。順道一提的是在蘇格蘭，凡是經過約150年以上時，就會自動登錄，或是不到一百五十年但卻有甚價值時，經政府的專家判定而也有可能收為保存對象。

在建築物被登錄在指定建築物中，或是相反地，被由其中被刪除時，都必須要告知持有人。但是，有一些以前就存在的情形，則有些持有人，並不被告知亦不知情。因為指定建築物相當多，所以希望那些住在古舊房屋裡的民眾，能向政府的設計部門加以詢問。因為和一般的建築物相較之下，指定建築物會有更多的嚴格限制。

指定建築物並不是就完全不能修建，只是必須要取得被稱為「指定建築物許可」的特別認可，當然只是重新粉刷油漆的普通作業，是不需要設

計許可的，但也有些情形是，僅只一點整修或內部改建，都需要設計許可或指定建築物許可。若是在沒有許可的情況下，自行修建時，除了會被命令回復原狀之外，還會被課以罰金，所以無論是作什麼樣的改建，還是事前和相關單位的設計人員商量一下比較好。

## 指定建築物的設計及許可的申請

關於指定建築物的設計申請及許可申請，是和前述普通的設計申請的手續相同的。是由國家（在蘇格蘭，就是由蘇格蘭開發部）和事前協議來決定的。有時會因情況需要而召開公聽會，由國家指定建築物的負責官員來聽取，出申請者會被要求口頭說明或是提出書面報告。

## 指定建築物持有人的義務

指定建築物的持有人有維持管理建築物，在必要時對建築物進行修繕的義務。如果因怠惰而使建築物呈現危險狀態時，持有人就會收到限時二個月內修繕的「修繕通知」。若是再置之不理時，地方或國家將會強制出售，若是建築物為空屋狀態時，國家或相關單位會代替持有人進行修繕，但是持有人必須支付修繕費用。這是在東羅吉安的手冊中提醒大家的。

## 補助金及貸款

因維持管理指定建築而需要額外的費用支出時，可以申請補助金或貸款。特別是建築物是十分傑出的作品時，政府機構（在蘇格蘭時，就是蘇格蘭歷史物築委員會）或是所屬的地方機構會給予補助金。但是在各地方的自治單位因經費預算不足時，就變成要自己負擔全部的費用了。在得到指定建築物許可的情形下進行之相關施工，可以減免付加價值稅（15%，現在是17%）。

## 保存地區的計劃

保存地區，地方自治的計劃當局或是在某些地方是國家（在蘇格蘭是蘇格蘭相關負責的大臣，英格蘭則是環境大臣），為了使一些具有歷史性建築的地方，在將來仍能繼續保存下來，而指定保存的。在保存地區為了維持地區全體之特色，所以在進行設計規劃上，較一般地區嚴格。

## 東羅吉安保存區

東羅吉安是個約有八萬人口的地方，東羅吉安的範圍內，有很多個指定保存區，合計共有23個地區。保存區的界線是在以前的聚落或村鎮中心區等，雖然是從仍保有過去傳統風格的屋舍集中區開始，但為了將周圍的農村景觀也能包含進

去，所以保存區的界線常常都是相當遼闊的。

## 在保存區的設計許可以及保存區許可

在保存區提出申請時，通常都用比平時更嚴格的標準來判定的，而且除此之外，還必須要有所謂的「保存區許可（CONSERCATON　AREA　CONSENT）」。下面就是受到比平時更嚴格管制的例子：

1.會對土地或建築物的外觀造成影響的（例如：窗櫺的更換，或是建築物顏色的改變等，可能會影響地保存區特色等）施工，都需要設計許可。

2.利用上的變更：在保存區內無論作何種用途變更（例如：住宅改為民宿，或是商店改營餐廳等等），都是需要設計許可的。

3.在建築用地內的小型施工：在自己家的小庭院內加蓋停車場，儲物處，或是門廊前的小停車區，圍牆，籬笆等，在一般地區不需要取得許可的小型施工，在保存區內卻都一定要先取得許可。

4.通常在拆除建築物時，並不需要許可，但在保存區時，就需要保存區許可。

5.屋外的廣告看板等：新設招牌或是更換舊有招牌時，即使是依照平常的規定，也是需要設計許可的。在英國，規定內除了最小型的廣告物之外，其餘的全都必須比照平時的設計許可和以同樣的程序來取得許可（而且包含期限）來辦理。

6.樹木：在保存區內的樹木，例如像是修剪枝椏等，會影響樹木外觀的任何行為，都必須在六個星期前知會設計規劃部門。

7.在保存區內的指定建築物的任何變動，都必須遵循指定建築物的相關設計法規。

在東羅吉安，被指定的23個保存區之一，卡瓦爾德村。

樹木的修剪整枝，就像是服裝飾品一般地在設計中，被廣為推展。在東羅吉安地區的所有樹木，全都是大而枝葉婆娑地保有其自然姿態。

## 保存區的設計方針

保存區的意義，不僅在於單純地維持現狀，還要能掌握機會（例如像是設計申請或是進行保存地區許可申請等）地創造改善，使之成為一個有更好環境的地方。

在東羅吉安地區，若有對保存區內設計許可和保存區許可的決定時，就會設置如下之方針。另外，因各保存區而有異，有時也會有更詳盡的個別適用方針。

(1) 在保存地區中進行「輪廓草圖（OUTLINE）許可」時，為了出示計劃案對周圍環境會有什麼影響，所以除非附有含周邊環境之草圖或是平面圖，否則不受理。

(2) 在進行詳細申請時，必須在一般的設計申請上，再加上詳細圖面，還必須附上可以清楚了解計劃和周邊環境關係的街道平面圖或是透視圖才可以。

(3) 保存區內所有的申請計劃，都必須在當地的報紙廣告欄上刊登公告。

(4) 關於建築物的拆除申請許可，僅限在於拆除有適切地重新修築的計劃時，才得以受理。

(5) 所有的申請案件，都必須考慮到其周圍的設計特色、大小、式樣、建築材料等，以求對保存街道景觀的貢獻。

(6) 保存區內大部份的建築物，個別價值都不如整體價值來得高。所以在提出申請時，也必須考慮到保存區內整體的價值。

(7) 窗戶、門、或者是屋頂的其他材料等的更換，或是屋頂窗（附加在屋頂的窗戶），玄關，門廊等施工時，或是建築物的加蓋等等，如對於設計不稍加注意，則就會損及保存區的特色。所以遇到這樣的作業申請時，就必須針對該建築物和其他周圍的建築物的調和與否作探討。

(8) 色彩是保存區景觀的重要要素之一。在進行建築物的粉刷時，必須要留意到房子本身的建築特色，例如屋簷或基礎線的帶狀紋，窗戶或是門框周圍的裝飾帶，或是塔樓等。在石頭上加以著彩，並不僅限於地方傳統，所以並不是一個適當的好方法。在建築物成群時，色彩的運用，應力求與周圍建築物相互調和。在色彩的調配和使用上，也可以請教政府相關部門設計部門，請求技術援助。

(9) 建築物一樓商店構造（SHOP FRONT）部份的設計，必須在和建築物全體有適切地相關性。不適切的建築材料或細部上的設計，也是不被許可的。

(10) 街燈、站牌、電話亭、郵筒、垃圾箱、交通號誌和招牌等，這些街道附屬設施和禁止停車的黃線等，都是構成街市重要部份。所以為了確保這些適切的設計設置位置，也必須得到設置這些設施的機構的許可。

(11) 在建築物上畫上或掛上招牌或廣告牌的大型設計，必須注意到是建築物整體的平衡。可以的話，是以採用簡單圖案的傳統設計為前提。內有燈管的電燈型招牌是決對不允許的。

(12) 對於保存區內的空屋或廢棄了的空屋，應積極地考量如何有效地利用。

(13) 樹木對於保存區的舒適性，有重大的意義，所以必須經由適切的管理來保護之。因此設計部門才會因此而對那些可能會修剪樹木的人提出了進言。

(14) 雖然因保存地區而異，但是有時也會對汽車的快捷舒適性造成不良的影響。

這時為了能將影響減至最輕，所以應該將交通管理方法列入，並尋求地方上的道路當區局的協助。

(15) 為了得到一般人和相關機構的協助，在保存地區實施的計劃，應該要以廣告或是海報方式來推動。

## 樹木的設計

樹木是襯托街道景觀的重要因素。所以無論是在哪一個國家，因為要保護現有的樹以免被砍伐，或是連根拔除以及切除上部枝幹等損害的行為，所以設置了很多規範和措施。例如在樹木被視為神聖義大利，即使是自家庭院內的樹，在嚴苛的規定上，不只是枯死，就是連移除，都是被視為是不可原諒的。即使是在英國，重要的樹木，或是成列的樹，或者是森林等，都有自治體所定的「樹木保護條例（TEEE PRESERVATION」，在條例中，對於受到保護的樹木，若是有任何可能會損及樹木的健康生長或形態的行為，樹木的所有者都必須於事前，取得自治體的許可。

在保存地區的樹木保護規定中，砍伐等作業，有規定必須在義務上，必須於事前通知；並且砍伐，有時也會被要求相對地植樹。在東羅吉安，必須於六週之前通知，在這期間，進行適當的應對之策的檢討，在必要時，也會將樹木保護條例適用於該樹等方法，以加強實行保護措施。

建築作業的計劃申請在進行檢討時，會針對現有的樹木植被和計劃的建築物的相關位置關係，或是和下水道的配管線路的關係等進行討論，有時也會因想留下樹木而指示變更計劃案。

不管樹木是否列入保護條例，若是缺乏慎重思考地隨意修整枝椏，就有可能會得到地方景觀變得十分難看。在東羅吉安，植樹或是修樹木時，都是由設計部門所屬的造園專家來推薦，繼砍倒的樹之後，應該要種植哪一種較為合宜等等，還可以給民眾這方面的建議。

● 表一　在進行樹木的修剪時…

◎原來的樹木

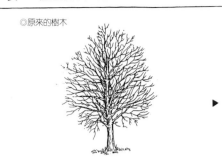

▼只是雜亂地將樹木的前端剪掉,這樣往後幾年內,
這棵樹都會以這樣醜陋的面貌存在。而且切口就這
麼地留在枝幹上,或許就會從切口處腐爛掉也說不定。

正確地選擇大枝葉,在樹幹的表面整齊地切口,這樣
切口才能塞住。樹木也才能夠保持其原來的風貌,又
吸收光線。

▼正確例子。

▼錯誤的修剪例

經過了兩年,雖然切口處都長出了新芽和嫩葉,但是
看起來就是十分脆弱,無法成為枝繁葉茂的大樹

出自:東羅吉安的計劃手冊。

# ●索引

## ●参考文献

Alsopp, Bruce 1985 : *Architecture in Britain and Europe*, Country Life Books, Felthum.
Aston, Michael and James Bond 1987 : *The Landscape of Towns*, Alan Sutton, Gloucester.
Auzas, Pierre-Marie 1979 : *Viollet le Duc 1814-1879*, Caisse National des Monuments Historique et des sites, Paris.
Bacon, Edmund N. 1974 : *Design of Cities (Revised Edition)*, The Viking Press, New York.
Bales, Mitzi (Ed.) 1988 : *Camping and Caravaning in Europe*, The Automobile Association, Basingstoke.
Bell, Simon 1993 : *Elements of Visual Design in the Landscape*, E & FN Spon, London.
Bingham, Heather 1988 : *A National Park in the Balance ?, GCSE Resource Guide 1*, Lake District National Park, Windermere.
Bingham, Heather 1988 : *Learning to Live With Tourism, GCSE Resource Guide 2*, Lake District National Park, Windermere.
Bresford, Maurice 1987 : *The Lost Villages of England*, Alan Sutton, Gloucester.
Bresford, Maurice 1988 : *New Towns of the Middle Ages*, Alan Sutton, Gloucester.
Bresford, Maurice and John G. Hurst (Ed.) 1989 : *Deserted Medieval Villages*, Alan Sutton, Gloucester.
Brook, Stephen and Charlie Waite 1986 : *The Dordogne*, George Philip, London.
Brunskill, R. W. 1971 : *Illustrated Handbook of Vernacular Architecture*, Faber and Faber, London.
Burke, Gerald 1976 : *Townscapes*, Penguin Books, Harmonsworth.
Caisse National du Credit Agricole,la 1985 : *Les Plus Beaux Villages de France*, Siege Social : Mairie de Collonges-La-Rouge.
Carter, Harold 1983 : *An Introduction to Urban Historical Geography*, Edward Arnold, London.
Champion, A.G. (Ed.) 1989 : *Counterurbanization*, Edward Arnold, London.
Chester, Carol 1986 : Holland, *Welcome to Belgium & Luxembourg*, Collins, Glasgow and London.
Clout, Hugh D. (Ed.) 1987 : *Regional Development in Western Europe, Third Edition*, David Fulton Publishers. London.
Clout, Hugh, Mark Blacksell, Russell King, David Pinder 1989 : *Western Europe, 2nd Edition*, Longman Scientific & Technical, Harlow.
Cook, Olive 1982 : *English Cottage and Farmhouses*, Thames and Hudson, New York
Crook, J. Mordaunt 1989 : *The Dilemma of Style*, John Murray, London.
Cullen, Gordon 1961 : *The Concise Townscape*, The Architectural Press, London.
Cullingworth, J. B. 1988 : *Town and Country Planning in Britain, Tenth Edition*, Unwin Hyman, London.
Darby, H. C. 1977 : *Domesday England*, Cambridge University Press, Cambridge.
Dixon, R. A. N. 1984 : *Welcome to Portugal*, Collins, Glasgow and London.
Dixon, R. A. N. 1985 : *Welcome to Spein*, Collins, Glasgow and London.
Dobby, Alan 1978 : *Conservation and Planning*, Hutchinson, London.
Dyson, Henry 1989 : *Your Home in France*, Longman, London.
East Lothian District Council 1988 : *East Lothian Planning Handbook*, Home Publishing Company Ltd, Wallington.
East Lothian District Council 1988 : *Dumbar Area Local Plan*, Huddington.
Eisenhardt, Jost 1987 : *The Visitor's Guide to Yugoslavia : the Adriatic Coast*, Moorland Publishing Co. Ltd, Ashbourne.
Finberg, Joscelyne 1987 : *Exploring Villages*, Alan sutton, Gloucester.
Findlay, Allan and Paul White (Ed.) 1986 : *West European Population Change*,Croom Helm, London.
Flower, John and Charlie Waite 1987 : *Provence*, George Philip, London.
Garner, J. F. and N. P. Gravells (Ed.) 1985 : *Planning Law in Western Europe, Second Edition*, North-Holland, Amsterdam.
German Commission for UNESCO 1975 : *The Renewal of Historic Town Centres in Nine European Countries*, Bonn.
German Commission for UNESCO 1980 : *Protection and Cultural Animation of Monuments, Sites and Historic Towns in Europe*, Bonn.
Girouard, Mark 1985 : *Cities and People*, Yale University Press, New Heaven.
Glyptis, Susan A. 1981 : *Leisure Life-styles*, Regional Studies Vol. 15, Pergamon Press, Oxford.
Grant, Michael 1979 : *The art and life of PONPEII and HERCULANEUM*, Newsweek Books, New York.
Hachette-guides Bleus 1988 : *Hachette Guide to Italy*, The Automobile Association, Basingstoke.
Hackney, Rod 1990 : *The Good, the Bad and the Ugly*, Frederick Muller, London.
Hall, Peter and Ann Markusen (Ed.) 1985 : *Silicon Landscapes*, Unwin Hyman, London.
Hammond, J. L. and Barbara Hammond 1987 : *The Village Labourer 1760-1832*, Alan Sutton, Gloucester.
Harrison, John and Shirley 1986 : *Welcome to Ireland*, Collins, Glasgow and London.
Harrison, John and Shirley 1987 : *Welcome to Austria & Switzerland*, Collins, Glasgow and London.
Hillman, Judy 1988 : *A New Look for London*, HMSO, London.
Hohenberg, Paul M. and Lynn Hollen Lees 1985 : *The Making of Urban Europe 100-1950*, Harvard University Press, Cambridge.
Holloway, J. Christopher 1989 : *The Bussiness of Tourism, Third Edition*, Pitman, London.
Hoskins, W. G. 1970 : *The Making of the English Landscape*, Penguin Books, Harmonsworth.
Humble, Richard 1989 : *Warfare in the Middle Ages*, Magna Books, Leicester.
Hutchinson, Maxwell 1989 : *The Prince of Wales : Right or Wrong ? An Architect Replies*, Faber and Faber, London.
Jencks, Charles 1988 : *The Prince the Architects*, Academy Editions, London.
Johnson-Marshall, Percy 1966 : *Rebuilding Cities*, Edinburgh University Press, Edinburgh.
Kain, Roger (Ed.) 1981 : *Planning for Conservation*, Mansell, London.
Keates, Jpnathan 1988 : *Tuscany*, George Philip, London.
King, Rebecca (Ed.) 1986 : *AA Where to go in Britain*, Automobile Association, Basingstoke.
Lawrence, C. H. 1989 : *Medieval Monasticism, Second Edition*, Longman, London.

Law, Bill 1991 : *Traditional Houses of Rural France*, Abbeville Press, New York.

Lowton, Richard (Ed.) 1989 : *The Rise and Fall of Great Cities*, Belhaven Press, London.

Maxwell, Flavia 1989 : *Your Home in Italy*, Longman, London.

Michelin Guide de Tourisme 1987 : *Alsace et Lorraine − Vosges*, Michelin et Cie, Clermont-Ferrand.

Michelin Tourist Guide 1985 : *French Riviera − Cote d'Azur*, Michelin et Cie, Clermont-Ferrand.

Michelin Tourist Guide 1985 : *Portugal*, Michelin et Cie, Clermont-Ferrand.

Michelin Tourist Guide 1986 : *Germany − West Germany and Berlin*, Michelin et Cie, Clermont-Ferrand.

Michelin Tourist Guide 1987 : *Dordogne − Perigord-Quercy*, Michelin et Cie, Clermont-Ferrand.

Michelin Tourist Guide 1987 : *Spein*, Michelin et Cie, Clermont-Ferrand.

Michelin Tourist Guide 1988 : *Italy*, Michelin et Cie, Clermont-Ferrand.

Michelin Tourist Guide 1989 : *Provence*, Michelin et Cie, Clermont-Ferrand.

Michelin Tourist Guide 1991 : *France*, Michelin et Cie, Clermont-Ferrand.

Morris, A.E.J. 1979 : *History of Urban Form, Second Edition*, Longman Scientific & Technical, Harlow.

Morrison, Robin and Christopher Fitz-Simon 1986 : *The Irish Village*, Thames and Hudson, London.

Moughtin, Cliff 1992 : *Urban Design : Street and Square*, Butterworth Architecture, Oxford.

Mumford, Lewis 1964 : *The Highway and the City (Revised Edition)*, Secker & Warburg, London.

Mumford, Lewis 1966 : *The City in History*, Penguin Books, Harmonsworth.

Mynors, Charles 1987 : *Planning Applications and Appeals*, Architectural Press, London.

Naismith, Robert J. 1989 : *The Story of Scottish Towns*, John donald Publishers, Edinburgh.

Nickels, Sylvie 1984 : *Welcome to Yugoslavia*, Collins, Glasgow and London.

OECD 1992 : *Tourism Policy and International Tourism in OECD Member Countries*, Paris.

Pearce, David 1989 : *Conservation Today*, Routledge, London.

Penoyre, John and Jane 1978 : *Houses in the Landscape*, Faber and Faber, London.

Postan, M. M. 1972 : *The Medieval Economy and Society*, Penguin Books. Hamondsworth.

Pothorn,Herbert : *Reizvolle deusche Kleinstadt*, Karl Muller Verlag Erlangen.

Prentice, Richard 1993 : *Tourism and Heritage Attractions*, Routledge, London.

Prince of Wales, the HRH 1989 : *A vision of Britain*, Doubleday, London.

Prizeman, John 1974 : *YOUR HOUSE the outside view*, Quiller Press, London

Quiney, Anthony and Robin Morrison 1987 : *The English Country Town*, Thames and Hudson, London.

Ravendale J.R. 1982 : *History on your Doorstep*, BBC, London.

Ray, Patrica 1978 : *Conservation Area Advisory Comittees*, Civic Trust, London.

Reader's Digest Association Ltd., the 1983 : *AA Book of British Villages*, Drive Publications Ltd., Basingstoke.

Reader's Digest Association Ltd., the 1984 : *Exploring Britain · Country Towns*, Reader's Digest, London.

Richards, Sir James et al. 1974-75 : *European Heritage, Issues One ~ Five*, Phebus Publishing Company, London.

Roberts, Brian K. 1982 : *Village Plans*, Shire Publications Ltd., Aylesbury.

Rougemont, Rosemary de 1989 : *Your Home in Portugal*, Longman, London.

Royal Borough of Kensington & Chelsea 1984 : *Urban Conservation & Historic Buildings*, Architectural Press.

Russell, Josiah Cox 1972 : *Medieval Regions and their Cities*, David & Charles, Newton Abbot.

Saalman, Howard 1968 : *Medieval Cities*, Studio Vista, London.

Saunders, Matthew 1987 : *The Historic Home Owner's Companion*, B. T. Batsford, London.

Scottish Civic Trust, the 1981 : *New Uses for Older Building in Scotland*, HMSO, Edinburgh.

Seaborne, Malcolm 1989 : *Celtic Crosses of Britain and Ireland*, Shire Publications Ltd., Aylesbury.

Sharp, Thomas 1940 : *Town Planning*, Penguin Books, Harmondsworth.

Sharp, Thomas 1946 : *The Anatomy of the Village*, Penguin Books, Harmondsworth.

Sharp, Thomas 1968 : *Town and Townscape*, John Murray, London

Shepherd J. and P. Gordon 1989 : *Small Town England*, Pergamon Press, Oxford.

Sullam, Joanna, Charlie Waite and John Ardagh 1988 : *Villages of France*, Weidenfeld and Nicolson, London.

Summerson, John 1963 : *Heavenly Mansions*, The Norton Library, New York.

Svenson, Per 1989 : *Your Home in Spain*, Longman London.

Tait, Joyce, Andrew Lane and Susan Carr 1988 : *Practical Conservation*, The Open University, Milton Keynes.

Tripp, H. Alker 1942 : *Town Planning and Road Traffic*, Edward Arnold & Co, London.

Tugnutt, Anthony and Mark Robertson 1987 : *Making Townscape*, Mitchell, London.

Waley, Daniel 1985 : *Later Medieval Europe, Second Edition*, Longman, London.

Woodward, G. E. O. 1975 : *The Dissolution of the Monastries*, Pitkin Pictorials Ltd., Andover.

Wright, H. Myles (Ed.) 1948 : *The Planner's Notebook*, The Architectural Press, London.

Wynn, Martin (Ed.) 1984 : *Planning and Urban Growth in Southern Europe*, Mansell, London.

Zucker, Paul 1959 : *Town and Square*, The M.I.T. Press, Cambridge.

## 感言

　　近年來在歐洲的各先進國家中，大都市的人口漸次減少，而另一方面，鄉間、村鎮的人口則隨之增加，產生了一種稱之為「逆都市化」的現象。而產生逆都市化現象的一個重要因素，就如同本書中所提及地，是為了適切地因應目前這個汽車化的社會，進而提高村鎮等的便捷及舒適。雖然如此，但是大部份的歐洲人為了找尋有歷史傳統之人性化居住環境，而從大都市遷移向城鎮村落的現象也是不容忽視的。

　　本書的目的是在於藉由許多照片來向讀者們介紹，在歐洲各國成功地根據歷史傳統而重建的人性化城鎮鄉落，以為參考範例。這些事例作為設計資料書籍，是為便利環境設計專家使用，因而依照設計的重要因素來做編排的，但這樣的編排結果，卻反而使得以表現傳統設計特色之地域性村落單位來看，環境設計的形態反而變得難以辨識。為了彌補這項缺點，期望各位能一併參閱前著之『歐洲的村落設計①、②』。而且為養成環境設計時，首尾一貫的感覺，與依據設計要素分別編排之本書相比，似乎以過去依國別，村落別區分之前著為佳。

　　不過，為針對依設計要素別編排之照片附加解說的本書，將歐洲優良之環境設計的共通特色，藉由簡明易懂的型式傳達給各位，相信亦有某種程度的成功。希望本書能獲一些對於有歷史傳統之人性化環境設計有興趣者的支持，更祈盼能對各位有所助益。

　　最後，在此特別向出版本書時，曾大力相助的畫報社編輯部長岡本義正先生，設計家小林茂男先生，及負責英文編寫部份的史考特・布勞斯先生，致上最誠摯的謝意。

## 井上　裕

　　1947年、出生於東京。畢業於東京大學工學院的建築系，是工學博士，也是一級檢定的建築師。1977—1978年，曾留學英國愛丁堡大學都市地域計畫學系。1988—1990年，任職於（財）國際開發中心，曾探訪、研究過亞洲、非洲、拉丁美洲等三十幾個國家。1988—1990年，曾任英國愛丁堡大學的伯特利克・蓋得斯計畫研究中心的客座研究員，並調查及研究歐洲諸國之鄉村及小城鎮之建設。亦曾任心瀉產業大學經濟學系教授，現在為海大學不動產學系的教授。曾有『歐洲之村落設計』①②等著作（畫報社發行）

　　現住址：〒239　神奈川縣橫須賀市野比3丁目24-2
　　FAX：0468-49-4141

# 北星信譽推薦・必備敎學好書

日本美術學員的最佳敎材

鉛筆畫技法　　粉彩筆畫技法　　沾水筆・彩色墨水技法　　野外寫生技法　　油畫質感表現技法

定價／350元　　定價／450元　　定價／450元　　定價／400元　　定價／450元

循序漸進的藝術學園；美術繪畫叢書

實用繪畫範本　　粉彩畫技法　　油畫基礎畫法　　水彩技法圖解

定價／450元　　定價／450元　　定價／450元　　定價／450元

最佳工具書

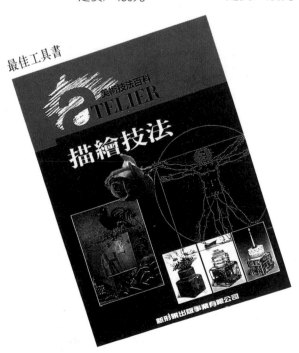

・本書內容有標準大綱編字、基礎素
　描構成、作品參考等三大類；並可
　銜接平面設計課程，是從事美術、
　設計類科學生最佳的工具書。
　編著／葉田園　　定價／350元

# 精緻手繪POP叢書目錄

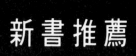

# 新形象出版圖書目錄

郵撥：0510716-5　陳偉賢　　TEL:9207133‧9278446　FAX:9290713　　地址：北縣中和市中和路322號8F之1

## 一、美術設計

| 代碼 | 書名 | 編著者 | 定價 |
|---|---|---|---|
| 1-01 | 新挿畫百科(上) | 新形象 | 400 |
| 1-02 | 新挿畫百科(下) | 新形象 | 400 |
| 1-03 | 平面海報設計專集 | 新形象 | 400 |
| 1-05 | 藝術‧設計的平面構成 | 新形象 | 380 |
| 1-06 | 世界名家挿畫專集 | 新形象 | 600 |
| 1-07 | 包裝結構設計 |  | 400 |
| 1-08 | 現代商品包裝設計 | 鄧成連 | 400 |
| 1-09 | 世界名家兒童挿畫專集 | 新形象 | 650 |
| 1-10 | 商業美術設計(平面應用篇) | 陳孝銘 | 450 |
| 1-11 | 廣告視覺媒體設計 | 謝蘭芬 | 400 |
| 1-15 | 應用美術‧設計 | 新形象 | 400 |
| 1-16 | 挿畫藝術設計 | 新形象 | 400 |
| 1-18 | 基礎造形 | 陳寬祐 | 400 |
| 1-19 | 產品與工業設計(1) | 吳志誠 | 600 |
| 1-20 | 產品與工業設計(2) | 吳志誠 | 600 |
| 1-21 | 商業電腦繪圖設計 | 吳志誠 | 500 |
| 1-22 | 商標造形創作 | 新形象 | 350 |
| 1-23 | 挿圖彙編(事物篇) | 新形象 | 380 |
| 1-24 | 挿圖彙編(交通工具篇) | 新形象 | 380 |
| 1-25 | 挿圖彙編(人物篇) | 新形象 | 380 |
|  |  |  |  |
|  |  |  |  |

## 二、POP廣告設計

| 代碼 | 書名 | 編著者 | 定價 |
|---|---|---|---|
| 2-01 | 精緻手繪POP廣告1 | 簡仁吉等 | 400 |
| 2-02 | 精緻手繪POP2 | 簡仁吉 | 400 |
| 2-03 | 精緻手繪POP字體3 | 簡仁吉 | 400 |
| 2-04 | 精緻手繪POP海報4 | 簡仁吉 | 400 |
| 2-05 | 精緻手繪POP展示5 | 簡仁吉 | 400 |
| 2-06 | 精緻手繪POP應用6 | 簡仁吉 | 400 |
| 2-07 | 精緻手繪POP變體字7 | 簡志哲等 | 400 |
| 2-08 | 精緻創意POP字體8 | 張麗琦等 | 400 |
| 2-09 | 精緻創意POP挿圖9 | 吳銘書等 | 400 |
| 2-10 | 精緻手繪POP畫典10 | 葉辰智等 | 400 |
| 2-11 | 精緻手繪POP個性字11 | 張麗琦等 | 400 |
| 2-12 | 精緻手繪POP校園篇12 | 林東海等 | 400 |
| 2-16 | 手繪POP的理論與實務 | 劉中興等 | 400 |
|  |  |  |  |
|  |  |  |  |

## 三、圖學、美術史

| 代碼 | 書名 | 編著者 | 定價 |
|---|---|---|---|
| 4-01 | 綜合圖學 | 王鍊登 | 250 |
| 4-02 | 製圖與議圖 | 李寬和 | 280 |
| 4-03 | 簡新透視圖學 | 廖有燦 | 300 |
| 4-04 | 基本透視實務技法 | 山城義彦 | 300 |
| 4-05 | 世界名家透視圖全集 | 新形象 | 600 |
| 4-06 | 西洋美術史(彩色版) | 新形象 | 300 |
| 4-07 | 名家的藝術思想 | 新形象 | 400 |
|  |  |  |  |

## 四、色彩配色

| 代碼 | 書名 | 編著者 | 定價 |
|---|---|---|---|
| 5-01 | 色彩計劃 | 賴一輝 | 350 |
| 5-02 | 色彩與配色(附原版色票) | 新形象 | 750 |
| 5-03 | 色彩與配色(彩色普級版) | 新形象 | 300 |
|  |  |  |  |
|  |  |  |  |

## 五、室內設計

| 代碼 | 書名 | 編著者 | 定價 |
|---|---|---|---|
| 3-01 | 室內設計用語彙編 | 周重彦 | 200 |
| 3-02 | 商店設計 | 郭敏俊 | 480 |
| 3-03 | 名家室內設計作品專集 | 新形象 | 600 |
| 3-04 | 室內設計製圖實務與圖例(精) | 彭維冠 | 650 |
| 3-05 | 室內設計製圖 | 宋玉眞 | 400 |
| 3-06 | 室內設計基本製圖 | 陳德貴 | 350 |
| 3-07 | 美國最新室內透視圖表現法1 | 羅啓敏 | 500 |
| 3-13 | 精緻室內設計 | 新形象 | 800 |
| 3-14 | 室內設計製圖實務(平) | 彭維冠 | 450 |
| 3-15 | 商店透視-麥克筆技法 | 小掠勇記夫 | 500 |
| 3-16 | 室內外空間透視表現法 | 許正孝 | 480 |
| 3-17 | 現代室內設計全集 | 新形象 | 400 |
| 3-18 | 室內設計配色手册 | 新形象 | 350 |
| 3-19 | 商店與餐廳室內透視 | 新形象 | 600 |
| 3-20 | 櫥窗設計與空間處理 | 新形象 | 1200 |
| 8-21 | 休閒俱樂部‧酒吧與舞台設計 | 新形象 | 1200 |
| 3-22 | 室內空間設計 | 新形象 | 500 |
| 3-23 | 櫥窗設計與空間處理(平) | 新形象 | 450 |
| 3-24 | 博物館&休閒公園展示設計 | 新形象 | 800 |
| 3-25 | 個性化室內設計精華 | 新形象 | 500 |
| 3-26 | 室內設計&空間運用 | 新形象 | 1000 |
| 3-27 | 萬國博覽會&展示會 | 新形象 | 1200 |
| 3-28 | 中西傢俱的淵源和探討 | 謝蘭芬 | 300 |
|  |  |  |  |
|  |  |  |  |

## 六、SP行銷‧企業識別設計

| 代碼 | 書名 | 編著者 | 定價 |
|---|---|---|---|
| 6-01 | 企業識別設計 | 東海‧麗琦 | 450 |
| 6-02 | 商業名片設計(一) | 林東海等 | 450 |
| 6-03 | 商業名片設計(二) | 張麗琦等 | 450 |
| 6-04 | 名家創意系列①識別設計 | 新形象 | 1200 |
|  |  |  |  |

## 七、造園景觀

| 代碼 | 書名 | 編著者 | 定價 |
|---|---|---|---|
| 7-01 | 造園景觀設計 | 新形象 | 1200 |
| 7-02 | 現代都市街道景觀設計 | 新形象 | 1200 |
| 7-03 | 都市水景設計之要素與概念 | 新形象 | 1200 |
| 7-04 | 都市造景設計原理及整體概念 | 新形象 | 1200 |
| 7-05 | 最新歐洲建築設計 | 石金城 | 1500 |

## 八、廣告設計、企劃

| 代碼 | 書名 | 編著者 | 定價 |
|---|---|---|---|
| 9-02 | CI與展示 | 吳江山 | 400 |
| 9-04 | 商標與CI | 新形象 | 400 |
| 9-05 | CI視覺設計(信封名片設計) | 李天來 | 400 |
| 9-06 | CI視覺設計(DM廣告型錄)(1) | 李天來 | 450 |
| 9-07 | CI視覺設計(包裝點線面)(1) | 李天來 | 450 |
| 9-08 | CI視覺設計(DM廣告型錄)(2) | 李天來 | 450 |
| 9-09 | CI視覺設計(企業名片吊卡廣告) | 李天來 | 450 |
| 9-10 | CI視覺設計(月曆PR設計) | 李天來 | 450 |
| 9-11 | 美工設計完稿技法 | 新形象 | 450 |
| 9-12 | 商業廣告印刷設計 | 陳穎彬 | 450 |
| 9-13 | 包裝設計點線面 | 新形象 | 450 |
| 9-14 | 平面廣告設計與編排 | 新形象 | 450 |
| 9-15 | CI戰略實務 | 陳木村 |  |
| 9-16 | 被遺忘的心形象 | 陳木村 | 150 |
| 9-17 | CI經營實務 | 陳木村 | 280 |
| 9-18 | 綜藝形象100序 | 陳木村 |  |

## 九、繪畫技法

| 代碼 | 書名 | 編著者 | 定價 |
|---|---|---|---|
| 8-01 | 基礎石膏素描 | 陳嘉仁 | 380 |
| 8-02 | 石膏素描技法專集 | 新形象 | 450 |
| 8-03 | 繪畫思想與造型理論 | 朴先圭 | 350 |
| 8-04 | 魏斯水彩畫專集 | 新形象 | 650 |
| 8-05 | 水彩靜物圖解 | 林振洋 | 380 |
| 8-06 | 油彩畫技法1 | 新形象 | 450 |
| 8-07 | 人物靜物的畫法2 | 新形象 | 450 |
| 8-08 | 風景表現技法3 | 新形象 | 450 |
| 8-09 | 石膏素描表現技法4 | 新形象 | 450 |
| 8-10 | 水彩・粉彩表現技法5 | 新形象 | 450 |
| 8-11 | 描繪技法6 | 葉田園 | 350 |
| 8-12 | 粉彩表現技法7 | 新形象 | 400 |
| 8-13 | 繪畫表現技法8 | 新形象 | 500 |
| 8-14 | 色鉛筆描繪技法9 | 新形象 | 400 |
| 8-15 | 油畫配色精要10 | 新形象 | 400 |
| 8-16 | 鉛筆技法11 | 新形象 | 350 |
| 8-17 | 基礎油畫12 | 新形象 | 450 |
| 8-18 | 世界名家水彩(1) | 新形象 | 650 |
| 8-19 | 世界水彩作品專集(2) | 新形象 | 650 |
| 8-20 | 名家水彩作品專集(3) | 新形象 | 650 |
| 8-21 | 世界名家水彩作品專集(4) | 新形象 | 650 |
| 8-22 | 世界名家水彩作品專集(5) | 新形象 | 650 |
| 8-23 | 壓克力畫技法 | 楊恩生 | 400 |
| 8-24 | 不透明水彩技法 | 楊恩生 | 400 |
| 8-25 | 新素描技法解說 | 新形象 | 350 |
| 8-26 | 畫鳥・話鳥 | 新形象 | 450 |
| 8-27 | 噴畫技法 | 新形象 | 550 |
| 8-28 | 藝用解剖學 | 新形象 | 350 |
| 8-30 | 彩色墨水畫技法 | 劉興治 | 400 |
| 8-31 | 中國畫技法 | 陳永浩 | 450 |
| 8-32 | 千嬌百態 | 新形象 | 450 |
| 8-33 | 世界名家油畫專集 | 新形象 | 650 |
| 8-34 | 插畫技法 | 劉芷芸等 | 450 |
| 8-35 | 實用繪畫範本 | 新形象 | 400 |
| 8-36 | 粉彩技法 | 新形象 | 400 |
| 8-37 | 油畫基礎畫 | 新形象 | 400 |

## 十、建築、房地產

| 代碼 | 書名 | 編著者 | 定價 |
|---|---|---|---|
| 10-06 | 美國房地產買賣投資 | 解時村 | 220 |
| 10-16 | 建築設計的表現 | 新形象 | 500 |
| 10-20 | 寫實建築表現技法 | 濱脇普作 | 400 |

## 十一、工藝

| 代碼 | 書名 | 編著者 | 定價 |
|---|---|---|---|
| 11-01 | 工藝概論 | 王銘顯 | 240 |
| 11-02 | 籐編工藝 | 龐玉華 | 240 |
| 11-03 | 皮雕技法的基礎與應用 | 蘇雅汾 | 450 |
| 11-04 | 皮雕藝術技法 | 新形象 | 400 |
| 11-05 | 工藝鑑賞 | 鐘義明 | 480 |
| 11-06 | 小石頭的動物世界 | 新形象 | 350 |
| 11-07 | 陶藝娃娃 | 新形象 | 280 |
| 11-08 | 木彫技法 | 新形象 | 300 |

## 十二、幼敎叢書

| 代碼 | 書名 | 編著者 | 定價 |
|---|---|---|---|
| 12-02 | 最新兒童繪畫指導 | 陳穎彬 | 400 |
| 12-03 | 童話圖案集 | 新形象 | 350 |
| 12-04 | 教室環境設計 | 新形象 | 350 |
| 12-05 | 敎具製作與應用 | 新形象 | 350 |

## 十三、攝影

| 代碼 | 書名 | 編著者 | 定價 |
|---|---|---|---|
| 13-01 | 世界名家攝影專集(1) | 新形象 | 650 |
| 13-02 | 繪之影 | 曾崇詠 | 420 |
| 13-03 | 世界自然花卉 | 新形象 | 400 |

## 十四、字體設計

| 代碼 | 書名 | 編著者 | 定價 |
|---|---|---|---|
| 14-01 | 阿拉伯數字設計專集 | 新形象 | 200 |
| 14-02 | 中國文字造形設計 | 新形象 | 250 |
| 14-03 | 英文字體造形設計 | 陳穎彬 | 350 |

## 十五、服裝設計

| 代碼 | 書名 | 編著者 | 定價 |
|---|---|---|---|
| 15-01 | 蕭本龍服裝畫(1) | 蕭本龍 | 400 |
| 15-02 | 蕭本龍服裝畫(2) | 蕭本龍 | 500 |
| 15-03 | 蕭本龍服裝畫(3) | 蕭本龍 | 500 |
| 15-04 | 世界傑出服裝畫家作品展 | 蕭本龍 | 400 |
| 15-05 | 名家服裝畫專集1 | 新形象 | 650 |
| 15-06 | 名家服裝畫專集2 | 新形象 | 650 |
| 15-07 | 基礎服裝畫 | 蔣愛華 | 350 |

## 十六、中國美術

| 代碼 | 書名 | 編著者 | 定價 |
|---|---|---|---|
| 16-01 | 中國名畫珍藏本 | | 1000 |
| 16-02 | 沒落的行業—木刻專輯 | 楊國斌 | 400 |
| 16-03 | 大陸美術學院素描選 | 凡谷 | 350 |
| 16-04 | 大陸版畫新作選 | 新形象 | 350 |
| 16-05 | 陳永浩彩墨畫集 | 陳永浩 | 650 |

## 十七、其他

| 代碼 | 書名 | 定價 |
|---|---|---|
| X0001 | 印刷設計圖案(人物篇) | 380 |
| X0002 | 印刷設計圖案(動物篇) | 380 |
| X0003 | 圖案設計(花木篇) | 350 |
| X0004 | 佐滕邦雄(動物描繪設計) | 450 |
| X0005 | 精細插畫設計 | 550 |
| X0006 | 透明水彩表現技法 | 450 |
| X0007 | 建築空間與景觀透視表現 | 500 |
| X0008 | 最新噴畫技法 | 500 |
| X0009 | 精緻手繪POP插圖(1) | 300 |
| X0010 | 精緻手繪POP插圖(2) | 250 |
| X0011 | 精細動物插畫設計 | 450 |
| X0012 | 海報編輯設計 | 450 |
| X0013 | 創意海報設計 | 450 |
| X0014 | 實用海報設計 | 450 |
| X0015 | 裝飾花邊圖案集成 | 380 |
| X0016 | 實用聖誕圖案集成 | 380 |

# 景觀設計實務

定價：850元

出版者：新形象出版事業有限公司

負責人：陳偉賢

地　　址：台北縣中和市中和路322號8Ｆ之1

門　　市：北星圖書事業股份有限公司

　　　　　永和市中正路498號

電　　話：9229000（代表）　ＦＡＸ：9229041

原　著：井上　裕

編譯者：新形象出版公司編輯部

發行人：顏義勇

總策劃：陳偉昭

文字編輯：雷開明

總代理：北星圖書事業股份有限公司

地　　址：台北縣永和市中正路462號5F

電　　話：9229000（代表）　ＦＡＸ：9229041

郵　　撥：0544500-7北星圖書帳戶

印刷所：皇甫彩藝印刷股份有限公司

行政院新聞局出版事業登記證／局版台業字第3928號
經濟部公司執／76建三辛字第21473號

國家圖書館出版品預行編目資料

景觀設計實務／井上裕原著；新形象出版公司
編輯部編譯. — 第一版. — 臺北縣中和市：
新形象，民86
　　面；　公分
參考書目：面
含索引
ISBN 957-9679-16-9(平裝)

1.景觀工程—設計

435.7　　　　　　　　　　　　　　　86002862